NOW PLAYING:

FROM SCRIPT TO SCREEN

FILM FUNDING PLAYBOOK– VOL 1

PUBLISHING
B|M
EST 2024

Published by: Bespoke Media Publishing

A Subsidiary of Bespoke Media Marketing LLC

All Bespoke Media Publishing Titles, Imprints, and Distributed Lines are available at special quantity discounts for bulk purchases for sales promotions, premiums, fund-raising, and educational or institutional use. Special book excerpts or customized printings can also be created to meet specific needs.

For details, contact the office of the Bespoke Media Publishing sales manager: info@bespokemediapublishing.com.

ISBN: 978-0-9858833-9-3

Unabridged Version First Printing: February 2025

10 9 8 7 6 5 4 3 2

Printed in the United States of America

DEDICATION

To my creations and my creators, my children—

You are the living proof of my dreams, the rhythm in my heartbeat, and the infinite inspiration behind every story I tell and every world I build. You are both the spark and the fire, the foundation and the future.

May you walk boldly in your own light, knowing that you carry within you the strength of generations, the magic of imagination, and the power to create beyond limits.

For all that you are and all that you will become—this is for you.

FOREWORD

Storytelling has always been the heartbeat of culture, shaping the way we see the world, ourselves, and each other. As technology evolves, so too does the way we tell and experience stories. Today, we stand at the intersection of cinema and Web3—a revolutionary moment where blockchain, decentralized communities, and digital ownership are redefining the way films are funded, distributed, and consumed.

This playbook is not just a guide—it is a manifesto for creators, investors, and visionaries who recognize the potential of Web3 in transforming independent film financing. It is an invitation to break free from traditional gatekeeping structures and embrace a decentralized future where artists have more control over their work, audiences become stakeholders, and stories find their most passionate champions.

As someone who has spent years at the forefront of entertainment, marketing, and emerging technology, I have witnessed firsthand the power of innovation in amplifying diverse voices and stories that deserve to be told. This playbook is built from experience, insights, and a deep belief that Web3 is not just a trend but an evolution of creative freedom and financial empowerment.

Whether you are a filmmaker looking to fund your next project, an investor seeking new opportunities in the digital landscape, or a studio executive exploring alternative distribution models, this book will equip you with the knowledge, strategies, and tools to navigate the Web3 film ecosystem.

The future of film is being written now. And in this new era, the power belongs to the creators and the communities that support them.

Welcome to the movement.

PREFACE
BY THE AUTHOR

For decades, independent filmmakers have struggled to break free from the constraints of traditional funding models—where studio gatekeepers, restrictive investors, and outdated distribution systems dictated who got a chance to tell their stories. As an independent creator, I have witnessed firsthand the financial barriers that prevent visionaries from bringing their ideas to life. But the landscape is changing, and a revolution is upon us.

I coined the term **GFT—Generational Financial Trauma** in 2021 to define the cycle of financial limitations that many creators, especially those from underrepresented backgrounds, have faced for generations. Access to capital, ownership of intellectual property, and fair revenue-sharing models have remained out of reach for far too long. **NFTs and blockchain technology provide the answer to breaking this cycle.**

With **NOW PLAYING – FROM SCRIPT TO SCREEN**, I aim to empower filmmakers, producers, and creatives with the knowledge and tools they need to leverage **Web3 technologies**—NFTs, DAOs, and decentralized finance—to fund and distribute their films on their own terms. **This is more than a playbook; it is a blueprint for creative sovereignty and financial independence.**

This book is not just about raising money for films; it's about redefining the economics of storytelling. It's about taking control of our narratives, owning our intellectual property, and ensuring that profits from creative works benefit the creators first.

If you've ever been told "no" by traditional financiers, if you've felt trapped in an industry that undervalues your work, or if you've been searching for a way to make your film dreams a reality **without sacrificing control**—this book is for you.

Welcome to the future of filmmaking. The power is now in your hands.

Nea Simone

Author

CONTENTS

THE NEW FRONTIER OF FILM FUNDRAISING

INTRODUCTION

The world of filmmaking has entered a transformative era, where innovation and technology intersect to empower creators in unprecedented ways. Web3 and Film3 technologies are not just buzzwords; they represent a paradigm shift in how films are financed, produced, and distributed.

In 2021, I originally coined the term **GFT—Generational Financial Trauma** to describe the systemic cycle of financial limitations that many creators, particularly those from underrepresented backgrounds, have faced for generations. It encapsulates the struggles filmmakers and artists endure when trying to secure funding, maintain ownership of their intellectual property, and achieve sustainable financial success in an industry historically controlled by a select few gatekeepers.

However, as Web3 technologies and blockchain-based financing models have gained traction, the acronym **GFT** has since been co-opted and repurposed to mean **Generative Film Technology** and **Generative Film Tokens.** These newer interpretations focus on using AI and blockchain to create, distribute, and monetize films through decentralized methods, often incorporating smart contracts and NFT-driven funding models. While these concepts align with the broader theme of financial empowerment and innovation in filmmaking, they diverge from the original intention of addressing economic inequities and systemic financial barriers for independent and marginalized filmmakers.

Despite its evolving usage, the core idea behind **Generational Financial Trauma** remains critical: Web3 is not just about technology—it's about **financial liberation** for artists. Whether it's through NFTs, DAOs, or alternative film funding models, the goal remains the same: to break cycles of financial exclusion and create a more equitable, creator-owned economy in the entertainment industry.

This book will guide you through this cutting-edge landscape, offering insights, strategies, and practical examples to help you navigate and thrive in the world of Film3. Together, we'll explore the tools and opportunities that are reshaping the industry—and empowering creators to take control of their vision.

UNDERSTANDING WEB3 AND FILM3

CHAPTER 1

Web3 refers to the third generation of the internet, emphasizing decentralization and user ownership through blockchain technology. Film3 applies these principles to the film industry, giving creators the tools to connect directly with audiences and maintain control over their projects. By integrating blockchain technology, decentralized platforms, and tokenomics, filmmakers now have the tools to break free from traditional gatekeepers and establish direct, meaningful connections with their audiences.

This new model allows creators to raise funds through community-driven initiatives, build loyal fanbases who actively support their projects, and explore innovative avenues for storytelling and revenue generation. Whether you're an independent filmmaker looking to finance your passion project or an established creator seeking new ways to engage your audience, understanding the Web3 ecosystem is essential.

Web3 and Film3 are revolutionizing the film industry by empowering creators to take control of their work, connect directly with audiences, and explore innovative funding and distribution methods. To fully leverage these tools, it's essential to understand the foundational concepts that underpin them.

This chapter introduces the basics of Web3 and its specialized subset, Film3, breaking down their key features and potential for transforming the filmmaking landscape.

What is Web3?

Web3 refers to the third iteration of the internet, characterized by decentralization, user ownership, and blockchain technology. i.e., Instead of relying on large corporations to manage data, Web3 enables individuals to control their data and assets, making it a natural fit for creative industries like film.

Example: Imagine a filmmaker using a platform where they own and control their content directly, without needing approval from a big studio or distributor.

What is Film3?

Film3 is a subset of Web3 specifically focused on transforming the filmmaking process. By using tools like non-fungible tokens (NFTs), decentralized autonomous organizations (DAOs), and smart contracts, Film3 enables creators to:

- Fundraise directly from their audience.
- Retain greater creative and financial control.
- Build loyal communities around their projects.

Example: A director sells NFTs of concept art to fans, raising funds while also giving those fans a sense of ownership in the film's journey.

Demystifying Web3 and Film3: A New Era for Filmmakers

The confluence of Web3 and Film3 is sparking an evolution in how stories are created, funded, and distributed. For filmmakers, the opportunity lies not just in leveraging blockchain technology, but in reshaping the very dynamics of the industry. While Web2 gave rise to streaming giants, centralizing content access, Web3 offers decentralization—giving creators direct control and a larger share of profits.

The primary value of Web3 for filmmakers is ownership. By tokenizing assets, creators can maintain rights over their work and ensure residuals through smart contracts. This stands in stark contrast to traditional distribution models where filmmakers often cede control to production houses or platforms. Film3 takes this a step further by creating ecosystems where audiences become co-creators and stakeholders.

For instance, projects like "The Infinite Machine," based on Ethereum's origin story, showcase how crowdfunding through NFTs can bring niche stories to life. These models empower filmmakers to connect with global audiences willing to invest not only money but their belief in the story's message. It's a radical shift that democratizes creative funding and brings filmmakers closer to their communities.

Moreover, decentralized autonomous organizations (DAOs) are offering filmmakers an innovative way to collaborate and share decision-making. DAOs provide an alternative to traditional studio hierarchies, allowing for collective governance in areas such as casting, funding, and profit sharing. As this ecosystem grows, filmmakers will have the tools to produce stories that reflect diverse voices without traditional gatekeeping.

Web3 and Film3 Explained: Unlocking the Future of Filmmaking

Understanding Web3 and Film3 is essential for filmmakers ready to embrace the digital frontier. Web3 represents the third generation of the internet, emphasizing decentralization through blockchain technology, while Film3 adapts these principles specifically for cinema and visual storytelling. But what does this mean for the filmmaker's toolkit?

One key insight lies in the advent of interoperable digital assets. Imagine a world where a filmmaker's NFT collection includes everything from concept art to final edits, easily transferable across platforms. This interoperability is not merely convenient; it redefines IP as portable and continuously monetizable. For example, creators can release limited-edition "director's cuts" as NFTs, offering die-hard fans exclusive content and securing additional revenue streams.

Another innovation is the real-time tracking of royalties. Blockchain's transparency ensures every stream, download, or ticket sale is logged immutably, and filmmakers are paid instantaneously through smart contracts. This eliminates the opacity of traditional revenue-sharing models where creators often wait months or years for earnings.

Film3's future also includes cross-media synergies. By integrating film projects with metaverse environments, filmmakers can craft interactive experiences that extend the story beyond the screen. For instance, a sci-fi film could include a virtual reality "training simulation" that immerses viewers in the protagonist's universe, creating deeper engagement and revenue opportunities.

The Web3 Revolution in Filmmaking: An Introduction to Film3

The revolution in filmmaking driven by Web3 and Film3 is redefining what it means to be a creator in the digital age. For filmmakers, the shift is as much about storytelling as it is about empowerment.

One groundbreaking opportunity in Film3 is fractionalized ownership. Through blockchain, filmmakers can sell "shares" of their projects, enabling micro-investors to fund films while earning dividends from future profits. This creates a more inclusive financing model where anyone can become a stakeholder in cinema's success.

In addition, Film3 introduces the concept of decentralized festivals. Traditional film festivals are expensive, competitive, and limited by geography. With Web3, filmmakers can host virtual festivals on decentralized platforms, showcasing their work to global audiences and bypassing the gatekeeping of traditional festival circuits. Attendees could even vote on awards using blockchain-based governance systems.

Finally, Film3 is democratizing access to creative resources. Decentralized networks connect filmmakers with global collaborators, from scriptwriters to VFX artists, without the barriers of location or industry elitism. This fosters an environment of borderless creativity where talent and ideas flow freely.

The Web3 revolution is more than a technological trend; it's a cultural shift. For filmmakers, it's an invitation to reimagine the possibilities of their craft and build a future where creativity knows no bounds.

KEY BENEFITS OF WEB3 FOR FILMMAKERS

CHAPTER 2

Web3 technology is a game-changer for filmmakers, providing tools to democratize funding, reach global audiences, and foster deeper fan engagement. By removing traditional barriers and creating new ways to interact with audiences, Web3 empowers filmmakers to achieve creative and financial independence.

This chapter explores the transformative benefits of Web3, with practical examples that demonstrate its potential for revolutionizing the film industry.

Democratized Funding

Filmmakers can bypass traditional gatekeepers by leveraging decentralized crowdfunding platforms and NFT sales to raise capital directly from their supporters.

Simple Example: A filmmaker creates a series of NFTs based on the film's storyboard, selling these digital collectibles to fund production without relying on studio backing.

Global Reach

Blockchain technology enables creators to tap into a global pool of investors and fans, breaking down geographical barriers.

Example: A filmmaker in Brazil uses an NFT marketplace to attract fans and investors from Japan, Germany, and the U.S., leveraging blockchain's international accessibility.

Fan Engagement

Web3 tools foster deeper audience interaction by offering fans unique benefits, such as ownership of exclusive content, voting rights, or direct involvement in the creative process.

Example: A filmmaker allows NFT holders to vote on the design of a key prop or costume, creating a sense of involvement in the creative process.

Exploring Web3 and Film3: The Foundations of Decentralized Cinema

Decentralized cinema, fueled by Web3 and Film3, is not just a buzzword; it's a paradigm shift toward a creator-first economy. Traditional filmmaking involves a labyrinth of middlemen—agents, studios, distributors—who dilute a creator's control and profits. With Web3, filmmakers can bypass these intermediaries, building direct pipelines to audiences.

One foundational principle of decentralized cinema is tokenized crowdfunding. Unlike platforms like Kickstarter, Web3 allows for direct, immutable transactions where supporters receive more than merchandise—they gain equity or benefits tied to the project's success. A real-world example is the film "Plantoid," an art project funded entirely through cryptocurrencies and governed by its community.

Film3 also introduces the concept of perpetual content engagement. A movie released under traditional models often stagnates after its initial run. However, with Film3, creators can mint "evolving NFTs" that unlock new scenes or spin-off content over time, ensuring long-term audience engagement.

Additionally, decentralized storage solutions like IPFS (InterPlanetary File System) ensure films are preserved outside the control of any single entity. This resilience safeguards against censorship or corporate takedowns, empowering filmmakers to push boundaries without fear of suppression.

Web3 Meets Film3: Understanding the Evolution of Storytelling

Storytelling has always evolved with technology, and the intersection of Web3 and Film3 is the latest chapter. From oral traditions to blockbuster franchises, the tools and frameworks of storytelling have shaped its reach and impact. In the Web3 era, storytelling becomes participatory and immersive.

Film3 empowers storytellers to invite audiences into the narrative creation process. Using NFTs, fans can vote on plot directions, character designs, or even casting choices. This participatory model fosters a deeper emotional investment, as audiences feel a sense of ownership in the story's evolution. It's a digital-age iteration of the "choose your own adventure" concept.

Moreover, blockchain technology provides unparalleled provenance. Creators can timestamp scripts, artwork, or production milestones on the blockchain, proving originality and protecting against plagiarism. This enhances trust in the creative ecosystem while also giving collectors unique opportunities to own verified pieces of cinematic history.

Web3's decentralized framework also transforms distribution. Filmmakers can bypass regional licensing restrictions, releasing their work globally without intermediaries. This global accessibility expands audience bases, especially for niche genres or culturally specific stories that traditional studios often overlook.

PREPARING FOR A SUCCESSFUL WEB3/FILM3 FILM FUNDRAISER

CHAPTER 3

Preparing for a Web3 fundraising campaign involves more than just understanding blockchain technology—it requires a solid foundation of community engagement, technical know-how, and a compelling pitch. By taking the time to educate yourself and your audience while building a dedicated supporter base, you can ensure your project's success in this new and exciting landscape.

This chapter provides practical steps and essential tips to help you lay the groundwork for a successful Web3 fundraiser.

Practical Steps and Essential Tips for a Successful Web3/Film3 Film Fundraiser

Fundraising in the Web3/Film3 ecosystem requires a strategic approach that leverages the unique benefits of blockchain technology. Here's a step-by-step guide to help filmmakers lay the groundwork for a successful campaign:

1. **Define Your Value Proposition**: Start by identifying what makes your film unique. Will you tokenize access, offer fractional ownership, or provide exclusive NFTs? Ensure your pitch highlights the benefits to supporters, such as financial returns, creative participation, or collectible value.

2. **Build Your Community Early**: Success in Web3 often hinges on a strong, engaged community. Leverage social platforms like Twitter, Discord, and Telegram to connect with crypto enthusiasts and potential backers. Share behind-the-scenes updates, concept art, and teasers to build anticipation.

3. **Choose the Right Blockchain**: Selecting a blockchain that aligns with your project's goals is crucial. Ethereum is widely used but comes with high fees; alternatives like Solana or Polygon may offer lower costs and faster transactions.

4. **Create and Mint NFTs**: Develop unique, high-quality NFTs tied to your film. These could include early-access passes, exclusive scenes, or even producer credits. Work with experienced designers to ensure your NFTs stand out in a competitive market.

5. **Launch a DAO for Governance**: Consider setting up a DAO to give supporters a voice in production decisions. Use tools like Aragon or Snapshot to establish transparent voting mechanisms that empower your community while keeping you in control.

6. **Promote Transparently**: Transparency builds trust. Use blockchain's inherent features to show how funds are being used. Regularly update backers on milestones achieved and the impact of their contributions.

7. **Collaborate with Influencers**: Partner with Web3 and Film3 influencers to amplify your campaign's reach. Their endorsement can attract a broader audience and lend credibility to your project.

8. **Host Virtual Events**: Organize live AMA (Ask Me Anything) sessions or virtual premieres to engage directly with your audience. Use these events to showcase your vision and answer questions about the fundraiser.

9. **Leverage Cross-Platform Integration**: Explore partnerships with metaverse platforms or gaming projects to enhance your film's visibility. This can add

an interactive layer to your campaign, attracting diverse backers.

10. **Set Realistic Goals**: Clearly outline your funding targets and break them into achievable milestones. Use tiered rewards to incentivize contributions at every level, from small investors to major backers.

By implementing these steps, filmmakers can harness the power of Web3 to democratize funding and create films that resonate with a global, digitally savvy audience.

FUNDRAISING STRATEGIES FOR FILM3

CHAPTER 4

Film3 offers a wealth of innovative fundraising strategies that empower filmmakers to raise capital while fostering deep connections with their audience. By leveraging NFTs, DAOs, and tokenomics, filmmakers can create unique and interactive opportunities for supporters to engage with their projects.

This chapter explores the key fundraising approaches in Film3, providing practical examples and insights into how each can be successfully implemented.

Below is an in-depth exploration of how creators, producers, and investors can raise funds in **Film3**—the emerging wave of filmmaking that leverages blockchain, NFTs, tokenization, and decentralized communities to finance and produce film projects. We will explore each approach, provide real-world or hypothetical examples, and offer practical tips for successful implementation.

1. NFT-Based Crowdfunding

Overview

NFTs (Non-Fungible Tokens) can serve as digital collectibles tied to a film project, representing fan memberships, digital art, exclusive behind-the-scenes content, or even partial rights to revenue streams. This approach allows filmmakers to raise capital by pre-selling unique digital items or perks directly to fans and investors.

How it Works

1. **Conceptualize the NFTs**: Determine the type of NFTs that will be minted—e.g., character art, concept art, storyboard stills, or "producer passes."
2. **Decide on Utility & Perks**: NFTs should have utility to incentivize investors. For instance, ownership could grant:
    - Access to special screenings or red-carpet events
    - Voting rights on creative decisions
    - Profit-sharing or royalty splits (where legally permissible)
    - Collectible items (e.g., limited-edition digital posters)
3. **Choose a Blockchain**: Ethereum, Polygon, and Solana are popular for NFT projects. Each has different minting costs and community sizes.
4. **Launch and Marketing**: Use social media, Web3 communities like Discord or Telegram, and partnerships with NFT influencers to build hype before release.
5. **Distribution of Funds**: The initial sale proceeds go to the production budget, and ongoing royalties (if programmed into the NFTs) can supplement future financing.

Example

- **Stoner Cats**: An animated series backed by NFTs sold to the community. Each NFT minted unlocked episodes of the show. This allowed the team to raise funds for production and engage fans who held the NFTs.

Tips for Success

- **Clearly outline benefits** to NFT holders, such as exclusive updates or behind-the-scenes materials.
- **Ensure transparency** around how raised funds will be spent.
- **Maintain momentum** by providing regular content and value to NFT holders—even after mint.

2. Tokenization and Fractional Ownership

Overview

Beyond simple collectibles, a filmmaker can tokenize an entire film project or aspects of it. Token holders can own a fraction of the intellectual property (IP), revenue streams, or other financial rights. This approach is more complex but potentially more lucrative and engaging for stakeholders.

How it Works

1. **Define the Asset**: Decide what is being tokenized—e.g., the film's revenue streams, distribution rights, or an entire production company's equity.
2. **Token Standard**: Use fungible tokens (like ERC-20 on Ethereum) to represent fractional ownership.

3. **Legal Structure**: Ensure regulatory compliance. Fractional ownership can be treated as a security, so engaging with legal experts is critical.
4. **Distribution Model**: Profits (e.g., box office, streaming licensing) are distributed proportionally to token holders.

Example

- 👌 **Republic's DAO Projects**: Through certain token-based fundraising platforms, accredited and retail investors purchase tokens that represent partial ownership or revenue rights in a project. As the film's revenue is generated, investors earn a return based on their token holdings.

Tips for Success

- ⊙ **Engage qualified legal counsel** from the beginning.
- ⊙ **Explain the financial model** in layman's terms to potential investors.
- ⊙ **Have a clear roadmap** for how investors see returns, including timelines and distribution methods.

3. Film/Production DAOs

Overview

A DAO (Decentralized Autonomous Organization) is an online community that pools funds, makes decisions collectively, and governs through blockchain-based voting. In Film3, DAOs can form around a shared cinematic vision, enabling contributors to co-create and co-fund projects in a democratic manner.

How it Works

1. **Formation & Charter**: Define the DAO's mission, governance rules, and funding processes.
2. **Raising Capital**: Members contribute crypto into a treasury in exchange for governance tokens.
3. **Voting & Decision-Making**: Proposals for script development, casting, or marketing are voted on by token holders.
4. **Profit-Sharing**: If the film generates revenue, the DAO manages distribution of profits to members per agreed-upon rules.

Example

- **Decentralized Pictures** (founded by Roman Coppola and others): While not a DAO in the strictest sense, it uses a token-based platform for filmmakers to pitch projects. The community votes on which projects receive grants, illustrating a decentralized approach to film financing.

Tips for Success

- **Craft a clear governance model**—emphasize simplicity to encourage widespread participation.
- **Foster community** through regular updates, AMA (Ask-Me-Anything) sessions, and interactive social media.
- **Establish transparent treasury management**—community members should see exactly where funds are allocated.

4. Traditional Crowdfunding with Web3 Add-Ons

Overview

Some filmmakers prefer a hybrid approach—using platforms like Kickstarter or Indiegogo (traditional crowdfunding) but layering on Web3 mechanics such as NFT rewards, blockchain-based fan clubs, or distribution of digital collectibles.

How it Works

1. **Set a Campaign Goal**: Similar to any crowdfunding project, define the funding goal and tiers of rewards.
2. **Integrate NFT Perks**: Offer NFT-based digital collectibles or tokens as one of the reward tiers.
3. **Educate Backers**: Provide tutorials on how to create a crypto wallet, claim NFTs, etc.
4. **Stretch Goals & Updates**: Offer additional NFT perks if the campaign exceeds certain milestones.

Example

- **A Short Film Project on Kickstarter**: The creator sets a $50,000 goal for production. Traditional backers can pledge money and receive typical rewards (DVDs, name in credits), but there are higher reward tiers where backers receive an NFT that grants them exclusive behind-the-scenes footage, or a future revenue share if certain conditions are met.

Tips for Success

- **Keep the barrier to entry low** for non-crypto natives: Simplify instructions on how to claim NFTs.
- **Leverage existing crowdfunding user bases** but emphasize Web3's unique benefits to stand out.
- **Build trust** by providing robust communication and meaningful updates throughout the campaign.

5. Specialized Film Web3 Funds and Grants

Overview

As Film3 grows, specialized funds and organizations (often structured as DAOs or blockchain-based ventures) emerge to finance Web3-based film projects. These funds may offer grants or investments in exchange for tokens or equity in the final production.

How it Works

1. **Research Potential Funds**: Identify Web3-focused venture capital firms, DAOs, or NFT-based film incubators.
2. **Pitch the Project**: Prepare a business plan, tokenomics overview, and examples of previous work.
3. **Negotiate Terms**: Terms might involve the fund acquiring tokens that represent a share of future revenue or rights to distribution.
4. **Leverage the Network**: Often, these specialized funds bring not just money but also strategic support, marketing help, and industry connections.

Example

- **Blockchain Creative Labs (by Fox Entertainment)**: Has invested in and incubated blockchain-based creative projects, including animated series. A producer with a Film3 idea could pitch to their incubator for a combination of funding and distribution partnership.

Tips for Success

- **Tailor your pitch** to show how Web3 integration benefits the fund's overall ecosystem.
- **Highlight unique selling points** such as community engagement or new revenue-sharing models.
- **Manage accountability** by providing milestone-based deliverables and transparent budgeting.

6. Hybrid Financing: Combining Traditional Investors with Web3

Overview

Not all film financiers are ready to dive headfirst into purely token-based deals. A hybrid model that combines traditional financing (equity, pre-sales, gap financing) with Web3 components (NFT perks, partial revenue tokenization, or DAO partnerships) can diversify a filmmaker's funding sources.

How it Works

1. **Secure Partial Traditional Funding**: For instance, approach film financiers, studios, or grant programs to cover a baseline of production costs.

2. **Supplement with Web3**: Launch NFTs, tokens, or a mini-DAO for additional capital.
3. **Allocate Returns**: Traditional investors may receive typical distribution returns, while NFT holders might get exclusive perks or fractional revenue from ancillary markets.
4. **Maintain Clarity**: Clearly define how different stakeholder groups are rewarded and which portions of the IP, revenue, or creative control they hold.

Example

- 👌 **Indie Film with Studio Support**: A small indie film secures 40% of its budget from a small production studio. To cover the remaining 60%, the producer mints NFTs that grant holders access to behind-the-scenes updates, potential cameo appearances, and a share of the film's streaming revenue.

Tips for Success

- **Document everything** so both traditional and Web3 investors understand their rights.
- **Use legal frameworks** that accommodate multiple investor classes.
- **Keep distribution strategies flexible** to account for both conventional and emerging channels.

7. Community-Driven Film Financing & Marketing

Overview

Building a strong community pre-production is crucial in Web3. Community-driven strategies mean your future audience can also be your financiers and marketing engine. Active Discord or Telegram groups can rally behind a film, generate buzz, and bring new backers.

How it Works

1. **Establish an Online Hub**: Whether it's Discord, Telegram, or a specialized Web3 platform.
2. **Incentivize Participation**: Offer tokens or NFTs to active community members who help with marketing, concept feedback, or storyline ideas.
3. **Regular AMAs and Polls**: Keep members engaged in the creative process. Community input can refine scripts or even casting decisions.
4. **Launch Micro-Funding Rounds**: Periodically release limited NFT collections or tokens tied to specific production milestones (e.g., costuming, set design, post-production).

Example

- **"Community First" Film**: A director shares weekly behind-the-scenes live streams. The community votes on minor plot points, and top participants receive NFT badges that could become collectible if the film succeeds.

Tips for Success

- **Be genuine**: Filmmaking requires leadership, so keep creative control while valuing community input.

- **Reward engagement**: Acknowledge community efforts with exclusive content or expanded roles in the project.
- **Plan for scaling**: As the community grows, so do management demands—moderators and administrators can help maintain order.

Film3 is revolutionizing the way films are financed by blending traditional methods with the power of blockchain, NFTs, and decentralized communities. From **NFT-based crowdfunding** to **tokenized ownership**, **DAOs**, **specialized Web3 funds**, and **hybrid models**, there is a broad spectrum of strategies that filmmakers can tailor to their specific needs and audiences.

Key Takeaways:

1. **Legal and regulatory compliance** is paramount when offering fractional ownership or profit-sharing tokens.
2. **Community building** is essential: Web3 thrives on engagement and transparency.
3. **Project clarity and consistent updates** can differentiate successful Film3 fundraising from failed campaigns.
4. **Hybrid approaches** can help bridge the gap between traditional investors and Web3 backers, maximizing exposure and resources.

By embracing these new fundraising approaches—and focusing on community involvement, innovative tokenomics, and transparent operations—filmmakers stand to gain not only capital but also a passionate fan base invested in the success of the final product.

GFTS – GENERATIVE FILM TOKENS AND GENERATIONAL FINANCIAL TRAUMA

CHAPTER 5:

The Evolution of Film Financing in Web3

The entertainment industry is undergoing a financial transformation, driven by blockchain technology and decentralized finance. One of the most significant innovations emerging from this shift is **Generative Film Tokens (GFTs)**—a blockchain-based funding model that challenges the traditional mechanisms of film financing. However, before **GFT** was widely recognized as an acronym for **Generative Film Token**, I originally coined the term **GFT – Generational Financial Trauma** in 2021 at the onset of NFT popularity.

At that time, **NFTs were being heralded as a breakthrough for artists**, offering them direct ownership and monetization of their work. Yet, while Web3 was creating new pathways for some, many underrepresented communities—particularly Black and brown creators—were still facing systemic financial exclusion. **GFT – Generational Financial Trauma** was a term I introduced to describe the **multi-generational cycle of economic disenfranchisement** that has plagued creative industries for decades, preventing artists from building sustainable wealth through their intellectual property.

Since then, the acronym **GFT** has been co-opted to mean **Generative Film Token**, shifting the focus toward a tokenized film funding model. However, both definitions remain deeply interconnected. The emergence of GFTs in film financing represents an opportunity to **break the cycle of generational financial trauma by decentralizing ownership and profits**, ensuring that wealth is created and retained within creative communities.

What Are Generative Film Tokens (GFTs)?

Generative Film Tokens are **blockchain-based assets** that allow investors, fans, and filmmakers to participate in the financing and revenue generation of a film project. Unlike traditional investment models, where major studios and hedge funds control the majority of profits, GFTs democratize access to film funding by enabling fractional ownership and **automated revenue distribution via smart contracts**.

These tokens offer multiple benefits, including:

- **Direct Participation in Film Revenues** – Token holders receive royalties from box office sales, streaming, licensing, and merchandising.
- **Decentralized Decision-Making** – Some GFT models allow investors and fans to vote on casting, production choices, and distribution strategies.
- **Access to Exclusive Content** – Holders may receive behind-the-scenes footage, VIP premieres, or digital collectibles tied to the film.

By leveraging Web3, **GFTs create a more equitable entertainment economy**, empowering independent filmmakers to bypass industry gatekeepers while providing financial stakeholders with transparent and fair revenue-sharing mechanisms.

GFT - Generational Financial Trauma: The Barriers That GFTs Must Address

While **Generative Film Tokens** introduce new financial opportunities, they must also be understood in the context of the original meaning of **GFT – Generational Financial Trauma**. This trauma has been perpetuated through systemic barriers that have historically prevented marginalized communities from accessing film financing, wealth-building opportunities, and long-term economic stability.

How Generational Financial Trauma Manifests in the Film Industry:

- **Lack of Funding Access** – Independent filmmakers of color, particularly Black and brown creatives, struggle to secure funding due to **institutional bias** in venture capital and Hollywood investment circles.
- **Predatory Contracts** – Historically, artists have been forced to sign **one-sided deals** that strip them of ownership in exchange for short-term financial support.
- **Exploitation of IP** – Many artists and filmmakers have had their creative work co-opted, rebranded, or outright stolen **without proper compensation**.
- **Wealth Drain** – Even when filmmakers find success, there are few financial structures in place to **ensure long-term sustainability** or allow for reinvestment in their communities.

By coining **GFT – Generational Financial Trauma** in 2021, I sought to highlight the **historical and ongoing financial oppression in creative industries**. The repurposing of the acronym into **Generative Film Tokens** presents an opportunity to **reclaim the financial narrative** and use technology to dismantle these systemic barriers.

Reclaiming Ownership Through GFTs

The rise of **Generative Film Tokens** represents an opportunity to **turn trauma into transformation**. By shifting film financing from an exclusionary model to a **decentralized, community-driven structure**, GFTs have the power to **correct historic injustices** and provide sustainable wealth-building opportunities for underrepresented creatives.

How GFTs Can Break the Cycle of Generational Financial Trauma:

- **Direct-to-Community Funding** – Independent filmmakers can fund projects without relying on major studios or biased financial institutions.
- **Equitable Revenue Distribution** – Smart contracts **automate fair payment structures**, ensuring that all contributors—filmmakers, actors, crew, and token holders—receive their rightful share of profits.
- **Long-Term Wealth Creation** – Filmmakers retain **ownership and IP rights**, allowing them to **pass down assets to future generations** rather than surrendering them to corporate interests.
- **Financial Literacy & Empowerment** – Educating creators and investors about **Web3 economics** ensures that marginalized communities benefit from this financial shift, rather than being left behind.

This model aligns with the **original intent behind GFT – Generational Financial Trauma**, serving as both a **corrective financial measure** and a **vehicle for economic justice** in the entertainment industry.

The Future of GFTs: Balancing Innovation with Equity

As **Generative Film Tokens** gain traction, it's critical to ensure that they are **used ethically and equitably**. While they present a radical new way to **fund and monetize films**, GFTs must be **structured with fairness, accessibility, and sustainability** in mind. Without safeguards, they risk becoming another tool for financial gatekeeping—this time within the Web3 space.

Moving forward, key considerations for ensuring **GFTs fulfill their potential** include:

- **Protecting Independent Creators** – Establishing fair legal frameworks that prevent exploitation of filmmakers.
- **Ensuring Inclusive Access** – Making GFT investments accessible to diverse communities rather than wealthy early adopters.
- **Educating the Next Generation** – Integrating Web3 financial literacy into film schools and creative incubators to **break cycles of economic disenfranchisement**.
- **Creating Community-Owned Platforms** – Developing decentralized distribution networks that allow filmmakers to monetize work **without relying on Big Tech or major studios**.

If implemented correctly, GFTs will **not only disrupt the financial structure of Hollywood** but also **correct decades of financial inequity**, proving that technology can be used as a force for **economic liberation**.

Restoring the Original Meaning of GFT

The **evolution of GFT from "Generational Financial Trauma" to "Generative Film Token"** is not just a linguistic shift—it is a reflection of the larger battle for **financial equity and creative autonomy**. While the term has been co-opted, the **original spirit**

behind GFT remains the same: to **break financial cycles that have historically kept independent filmmakers from building generational wealth**.

Through blockchain technology and decentralized ownership, we now have the tools to **reclaim what was taken, rewrite the rules, and ensure that our stories—and the wealth they generate—remain in our hands**. GFTs are more than a financial mechanism; they are a **movement toward justice, equity, and long-term economic empowerment**.

The question isn't just **how** GFTs will reshape the industry—it's **who will control them**. And the answer must be: **the creators, the communities, and the generations yet to come**.

REAL-WORLD APPLICATIONS OF GFTS – CASE STUDIES IN DECENTRALIZED FILM FUNDING

CHAPTER 6:

As the film industry embraces Web3 technology, **Generative Film Tokens (GFTs)** are proving to be more than just a financial tool—they are transforming the way films are funded, produced, and distributed. By decentralizing the financial structure of filmmaking, GFTs empower creators, engage communities, and redefine audience participation.

This chapter explores real-world applications of **GFTs and decentralized funding models**, showcasing how filmmakers and production companies are leveraging blockchain technology to **democratize access to film investment and create sustainable revenue streams**.

Overview:

Based on the book *The Infinite Machine* by Camila Russo, this film tells the story of Ethereum's creation and its impact on the world of decentralized finance. Rather than relying on traditional studio funding, the producers launched an NFT collection to raise capital.

How GFTs Played a Role:

- The project sold **10,499 NFTs**, raising over **$2.5 million** to fund the film's production.
- NFT holders were granted **exclusive access to behind-the-scenes content**, merchandise, and voting rights on creative decisions.

- The funding model allowed the production team to retain **full creative and financial control**, eliminating the need for corporate investors.

Why It Matters:

- Demonstrated that blockchain communities **can finance large-scale film productions** without relying on traditional Hollywood financing.
- Created a **new revenue model where NFT holders become active stakeholders** in the film's success.

Case Study #2: "Calladita" - The First DAO-Funded Indie Film

Overview:

Directed by Miguel Faus, *Calladita* became the **first feature film fully funded by a decentralized autonomous organization (DAO)**. This Spanish indie film raised money through **NFT sales** and received backing from a film-specific DAO.

How GFTs Played a Role:

- The team **minted NFTs that offered supporters exclusive access to the film's development** and behind-the-scenes content.
- Investors **received a stake in the film's revenue** and were granted **creative input through the DAO structure**.
- The project secured a **grant from the Decentralized Pictures Foundation (co-founded by Roman Coppola)**, reinforcing the legitimacy of blockchain funding in filmmaking.

Why It Matters:

- Marked a **paradigm shift** in indie filmmaking, proving that DAOs and GFTs can **replace traditional investors**.
- Showcased how **community-backed funding empowers independent filmmakers**, bypassing Hollywood's traditional financial barriers.

Case Study #3: "Hollywood DAO" - A Web3 Studio for Decentralized Film Production

Overview:

Hollywood DAO is a **decentralized collective aimed at funding and producing films** outside of the studio system. This Web3 initiative allows filmmakers to pitch ideas directly to a community of investors who vote on which projects receive funding.

How GFTs Played a Role:

- The DAO sells **GFTs that represent fractional ownership in film projects**, distributing profits directly to token holders.
- Investors participate in **governance decisions**, including casting choices, marketing strategies, and distribution deals.
- The DAO partnered with **Netflix and other streaming platforms** to bring Web3-funded films to mainstream audiences.

Why It Matters:

- Introduced a **community-driven studio model**, where film financing and decision-making are **decentralized**.
- Proved that blockchain funding **can coexist with mainstream entertainment platforms**, bridging the gap between Web3 and Hollywood.

Case Study #4: "Flinch" - The First Film Sold as an NFT

Overview:

Flinch, an independent crime thriller, made history as **the first feature film sold as an NFT**. The film's rights were auctioned on the blockchain, allowing buyers to own **distribution and licensing rights** through smart contracts.

How GFTs Played a Role:

- Instead of selling the film's rights to a major distributor, the producers auctioned an **NFT that granted ownership over the film's revenue streams**.
- The buyer received **royalties from streaming deals, digital downloads, and merchandising**.
- The NFT sale **created a direct financial connection between the buyer and the film's long-term profitability**.

Why It Matters:

- Redefined **film distribution**, showing that **NFTs can be used to sell film rights and bypass traditional distribution models**.

- Created a **direct-to-consumer ownership model**, where investors become **true stakeholders in the film's success**.

Case Study #5: "Antara" - A $50 Million Film Funded Through NFTs

Overview: *Antara* is an ambitious Hollywood epic that tells the true story of Antara ibn Shaddad, a Black slave who won his freedom and became a knight and renowned poet in ancient Arabia. The project aimed to raise $50 million through the sale of NFTs, allowing the public to own a stake in the film's rights and earn a yield during its production phase. The film was set to begin production in the Neom desert in Saudi Arabia.

- paus.tv

How GFTs Played a Role:

- **Community Engagement:** The project launched the "Arabian Camels" NFT collection, granting holders exclusive benefits such as set visits, acting roles, and premiere tickets.
- **Decentralized Funding:** By offering fractional ownership through NFTs, the film democratized investment opportunities, enabling a global audience to participate in its financing.

Why It Matters:

- **High-Value NFT Funding:** *Antara* set a precedent for large-scale film financing through NFTs, showcasing the potential for significant capital raises outside traditional channels.
- **Empowering Diverse Narratives:** The project highlights how decentralized funding can bring culturally rich and

diverse stories to the forefront, bypassing conventional industry gatekeepers.

Case Study #6: "KillRoy Was Here" - Kevin Smith's NFT Movie Release

Overview: Directed by Kevin Smith, *KillRoy Was Here* is a horror-comedy anthology that made history by being released exclusively as a series of NFTs. In July 2022, 5,555 unique tokens were minted, each granting the owner access to the film and potential participation in future content creation.

- en.wikipedia.org

How GFTs Played a Role:

- **Exclusive Access:** NFT holders received the full movie and were given rights to create and profit from their own derivative works using characters from the film.
- **Community Collaboration:** This model fostered a collaborative creative environment, where fans could contribute to the film's expanding universe.

Why It Matters:

- **Innovative Distribution:** By releasing *KillRoy Was Here* as an NFT, Kevin Smith bypassed traditional distribution channels, retaining creative control and directly engaging with his audience.
- **Fan Empowerment:** The approach empowered fans to become co-creators, blurring the lines between content consumers and producers.

Case Study #7: "No Postage Necessary" - The First Blockchain-Based Film Release

Overview: *No Postage Necessary* is a romantic comedy-drama that became the first film to be released via blockchain technology. In July 2018, the film was made available for streaming using cryptocurrency payments through a decentralized application, marking a significant milestone in film distribution.

- en.wikipedia.org

How GFTs Played a Role:

- **Blockchain Distribution:** The film was distributed on the Vevue platform, allowing viewers to rent or purchase the movie using cryptocurrency, ensuring secure and transparent transactions.
- **Decentralized Access:** This method provided an alternative to traditional streaming services, reducing intermediary costs and enhancing revenue for creators.

Why It Matters:

- **Pioneering Digital Release:** *No Postage Necessary* showcased the viability of blockchain as a distribution method, paving the way for future projects to explore decentralized platforms.
- **Financial Innovation:** The film's release demonstrated how cryptocurrencies could be integrated into mainstream media consumption, offering new financial models for the industry.

These case studies exemplify the transformative potential of GFTs and decentralized funding in the film industry. By embracing blockchain technology, filmmakers can explore innovative financing and distribution models, fostering direct engagement with audiences and redefining traditional industry paradigms.

THE FUTURE OF GFTS IN FILM FINANCE: THE NEXT FRONTIER IN ENTERTAINMENT

CHAPTER 7

The emergence of Generative Film Tokens (GFTs) marks a fundamental shift in film financing, granting independent filmmakers unprecedented financial autonomy, audience engagement, and long-term wealth-building opportunities. As Web3 technology continues to evolve, these tokens will redefine traditional funding structures, ensuring a more inclusive, transparent, and equitable entertainment ecosystem.

By leveraging blockchain and decentralized finance (DeFi), filmmakers now have the ability to:

1. Bypass Gatekeepers and Finance Projects Through Web3 Communities

For decades, the Hollywood studio system has dictated which films get made, who gets to tell their stories, and who reaps the financial benefits. Independent filmmakers have often struggled to break through this system due to a lack of funding and access to distribution channels.

GFTs disrupt this model by allowing filmmakers to raise funds directly from their communities.

- Instead of pitching to traditional studios or financiers, filmmakers can mint tokens representing shares in their project, selling them directly to fans, investors, and supporters.
- Platforms like Decentralized Autonomous Organizations (DAOs) create crowdfunded film studios, where token holders collectively decide on funding and production decisions.

- This direct-to-investor approach eliminates reliance on traditional middlemen (executives, producers, distributors) who often shape content according to their commercial interests, rather than artistic integrity.

With GFTs, the power shifts back to the creators, who can now develop projects based on audience demand and community support rather than studio politics.

2. Ensure Fair Revenue Distribution Using Smart Contracts

One of the biggest pain points in the film industry has been unequal revenue sharing. Studios, distributors, and financiers often take the lion's share of profits, leaving independent filmmakers with little financial return—even on successful projects.

With GFTs and smart contracts, revenue distribution can be automated, transparent, and immutable:

- Smart contracts ensure instant and fair payments to token holders based on pre-agreed terms, eliminating disputes over residuals and backend profits.
- Cast and crew members can receive micro-payments via tokenized contracts, ensuring that every stakeholder gets their fair share of revenue.
- Token holders can earn passive income from their investment in a film's success, providing long-term financial benefits beyond traditional box office or streaming deals.

This model eliminates financial gatekeeping, ensuring equitable earnings for creators, investors, and backers while also reducing fraud and mismanagement in film financing.

3. Create Deeper Audience Engagement, Where Fans Become Financial Stakeholders

The traditional film industry treats audiences as passive consumers, offering them no real stake in the success of a film. However, with GFTs, fans become active financial stakeholders, meaning they have a vested interest in a project's success.

How does this change the game?

- Token holders get exclusive benefits, such as early access to screenings, behind-the-scenes content, and VIP experiences.
- Fans can participate in creative decision-making, such as voting on casting, marketing strategies, or alternate endings.
- Gamification elements—such as unlocking special content based on token ownership—make the film experience more interactive.

This new level of engagement fosters a deeper connection between creators and their audiences, ensuring stronger financial backing, organic marketing, and long-term loyalty.

4. Retain Ownership and Long-Term Profit Potential

In the traditional industry, independent filmmakers are often forced to sell off distribution rights or sign unfair deals just to get their projects made. This results in losing control over their intellectual property (IP) and missing out on long-term revenue.

With GFTs, creators retain full ownership of their work, allowing them to:

- Monetize their film repeatedly through tokenized licensing, merchandising, and streaming rights.
- Generate passive income by keeping a percentage of token resale revenue (similar to NFT royalties).
- Control creative and distribution strategies, ensuring their projects reach the right audiences without compromising their vision.

By leveraging blockchain, filmmakers can turn their projects into long-term financial assets rather than one-time paychecks.

Challenges and Considerations for Widespread Adoption

While the future of GFTs in film finance is promising, several obstacles must be addressed before this model becomes mainstream.

1. Regulatory Uncertainty

- Governments and financial institutions are still catching up with blockchain-based securities, leading to evolving regulations around tokenized investments.
- SEC regulations in the U.S. and international compliance laws could impact how GFTs are sold, traded, and structured.
- Clearer legal frameworks are needed to prevent fraud, protect investors, and ensure that filmmakers comply with security laws while tokenizing their projects.

2. Market Volatility

- The value of GFTs and NFT-backed film projects can fluctuate dramatically based on market trends, potentially impacting investor confidence.
- A volatile crypto market could lead to speculative investment behavior, creating financial risks for both creators and investors.
- Filmmakers must structure their tokenomics carefully, ensuring that tokens are tied to real-world utility and sustainable revenue streams rather than hype.

3. Education and Adoption

- Many filmmakers, producers, and investors still lack technical knowledge about Web3 financing, making adoption slow.
- Studios and industry professionals are hesitant to embrace decentralization, fearing loss of control over traditional revenue models.
- Educational initiatives and partnerships with blockchain experts will be critical in making Web3 funding accessible, scalable, and sustainable.

What's Next for GFTs in Film Finance?

As more filmmakers experiment with GFTs, DAOs, and blockchain-based financing, we can expect major shifts in how films are funded, produced, and monetized.

1. More Mainstream Adoption by Major Studios and Streaming Platforms

- Traditional studios and streaming giants like Netflix, Amazon, and Disney will integrate GFTs into their funding models, using blockchain for fairer revenue distribution.

- Hybrid Web2/Web3 models will emerge, where films are partially funded through GFTs while maintaining traditional studio backing.
- Celebrity-led GFT projects will increase adoption, as more A-list actors and directors recognize the benefits of direct funding and audience ownership.

2. Stronger Legal Frameworks to Protect Tokenized Investments

- New SEC guidelines and global regulations will clarify how GFTs can be issued, traded, and managed, increasing investor confidence.
- Intellectual property laws will evolve to recognize tokenized film assets, preventing exploitation or piracy of blockchain-funded projects.
- Industry-specific Web3 financing platforms will emerge, offering legal and compliance support for filmmakers.

3. New Hybrid Funding Models

- Blended financing approaches will merge traditional studio funding, private equity, and decentralized film tokens, providing greater financial flexibility for creators.
- Tokenized distribution networks will allow films to be released via blockchain-based streaming services, ensuring direct-to-audience monetization.

- Secondary markets for GFTs will grow, allowing investors to trade tokens post-release, creating a new ecosystem of film-based financial assets.

GFTs as a Catalyst for Economic Transformation

GFTs are not just a technological innovation—they are a revolutionary shift in financial empowerment for filmmakers. By decentralizing power, these models correct historic financial disparities and give independent creatives an unprecedented level of control over their work.

The transition from traditional Hollywood gatekeeping to Web3 decentralization represents one of the most significant disruptions in film history. The question now is who will embrace this transformation and lead the charge into the future of film finance?

The answer lies in those who are willing to learn, adapt, and innovate—because the future of entertainment is being written on the blockchain.

The revolution is already here. The only question is: **Who will take advantage of it first?**

GFTs have already proven their viability in film financing, and as Web3 adoption grows, these models will become **a mainstream alternative to Hollywood's traditional funding structures**. Whether through **NFT sales, DAOs, or tokenized revenue sharing**, blockchain technology is reshaping **who gets to tell stories, how they get funded, and who profits from them**.

PREDICTIONS: THE FUTURE OF GFTS AND DECENTRALIZED FILM FINANCE

CHAPTER 8

As **Generative Film Tokens (GFTs) and decentralized funding models** gain momentum, they will reshape the entertainment industry in profound ways. The following predictions highlight how this financial revolution will unfold over the next **five to ten years**, setting new standards for funding, distribution, and audience engagement.

1. Traditional Studios and Streaming Giants Will Adopt GFTs

While independent filmmakers are leading the charge, **major studios and streaming platforms will inevitably enter the space** to remain competitive.

Netflix, Amazon Prime, Disney+, and HBO Max will introduce hybrid Web2/Web3 funding models. Studios will tokenize a percentage of their projects, allowing fans to invest in major franchises. This will create a new **direct-to-fan ownership model**, similar to how sports teams are selling fan tokens today.

Tokenized revenue-sharing models will become common. Instead of relying solely on streaming subscriptions, platforms will experiment with **pay-per-view blockchain contracts**, ensuring that token holders receive a percentage of streaming profits.

Hollywood talent will demand GFT-backed contracts. Major actors, directors, and producers will negotiate deals that **include tokenized profit-sharing agreements**, similar to the way Robert Downey Jr. negotiated backend points for *Avengers: Endgame*.

2. Film DAOs Will Compete with Traditional Studios for Market Share

The rise of **Decentralized Autonomous Organizations (DAOs) focused on film production** will **challenge the dominance of Hollywood studios** by offering fairer, community-driven funding models.

Community-funded movies will rival traditional blockbusters. Web3-native films, funded through DAOs and GFTs, will secure **A-list talent** and **compete at major film festivals like Sundance and Cannes**.

Fan-driven film studios will emerge. Decentralized collectives will fund, vote on, and greenlight movies, **disrupting the traditional development pipeline**. Think of this as a **Reddit-style voting system for film production**, but with financial stakes.

Studios will acquire DAOs or partner with them. As DAOs gain influence, legacy studios will **either invest in them or acquire them outright**, similar to how Netflix has absorbed independent production companies.

3. GFTs Will Create a Secondary Market for Film-Based Financial Assets

Right now, when a film makes money, profits go to studios and distributors. But **what if movie tokens could be bought, sold, and traded like stocks?**

GFT exchanges will emerge. Investors will **trade tokens representing stakes in films**, allowing them to profit from box office and streaming success **without waiting for traditional revenue cycles**.

Tokenized pre-sales will replace risky box office predictions. Instead of relying on industry analysts, studios will **track token trading volume** to predict a film's commercial success **before release**.

Filmmakers will retain long-term passive income. Instead of selling off rights, creators will **earn perpetual royalties from token resales**, ensuring **lifelong financial stability**.

4. The Rise of Blockchain-Based Film Festivals and Tokenized Voting

Film festivals and award shows will evolve to **embrace Web3 voting, distribution, and monetization models.**

Web3-native film festivals will emerge. New festivals will screen exclusively **blockchain-funded films**, offering **tokenized ticketing and voting rights** to attendees.

Film voting will become decentralized. Awards like the **Oscars and Cannes Palme d'Or** may incorporate **blockchain-based voting systems**, allowing audiences to participate in **transparent, tamper-proof selections**.

Tokenized film premieres will create global access. Instead of limited festival screenings, films will **launch as token-gated streaming events**, allowing anyone with GFTs to attend an exclusive first look.

5. AI and Smart Contracts Will Automate Film Revenue Distribution

The integration of **AI-driven finance models and blockchain technology** will remove many inefficiencies in film financing and royalty payments.

Instant, automated payouts will become the norm. AI-driven smart contracts will **calculate and distribute revenues in real time**, ensuring that every stakeholder—from actors to investors—**receives payments instantly** as revenue flows in.

AI will predict film funding success. Filmmakers will use **AI-driven analytics** to determine whether a GFT-backed film has **strong investor interest before it even launches**.

Dynamic revenue sharing will adjust in real time. If a film **outperforms box office expectations**, smart contracts will **automatically adjust revenue shares**, ensuring equitable profit distribution among investors and creators.

Final Prediction: GFTs Will Be the Standard for Independent Film Finance

Within the next **5-10 years**, it will become **industry standard for independent films to be financed through GFTs and DAOs**.

Crowdfunding via Kickstarter and GoFundMe will be replaced by GFT models. Instead of donating, backers will **own a financial stake in projects** with **built-in revenue-sharing mechanisms**.

Independent filmmakers will use tokenized IP to retain ownership. Filmmakers will **license and monetize their work on blockchain platforms**, ensuring **continuous income generation** instead of one-time payouts.

The power dynamic in Hollywood will shift. Traditional gatekeepers will lose influence as more creators turn to **decentralized funding and distribution models**.

The Film Industry's Blockchain-Backed Future

The next decade will define whether **GFTs and Web3 funding models** truly transform film finance or remain a niche movement. If the **current momentum continues**, we will witness:

A new era of independent filmmaking—where creatives are **financially empowered** and **control their IP**.

A film industry where audience investment and participation drive success, instead of outdated studio models.

A shift from centralized decision-making to decentralized, community-driven filmmaking—a true **democratization of entertainment**.

As this revolution unfolds, the **biggest question remains**: **Will the Hollywood elite adapt, or will they be left behind?** The future of film finance is being written **on the blockchain**, and those who move first will shape the next golden age of cinema.

MITIGATING RISKS IN THE FUTURE OF GFTS AND DECENTRALIZED FILM FINANCE

CHAPTER 9

While **Generative Film Tokens (GFTs)** and decentralized funding models offer groundbreaking opportunities, they also come with **significant risks**. If these challenges are not **carefully addressed**, the Web3 film revolution could face regulatory roadblocks, financial instability, or even exploitation by bad actors. Below are the **major risks** associated with GFT-backed film financing and **proven strategies to mitigate them.**

1. Regulatory Uncertainty - Navigating Legal and Compliance Issues

The Risk:

- Governments and financial regulators **are still catching up with blockchain-based securities**, creating legal uncertainty.
- **GFTs may be classified as securities**, subjecting filmmakers to strict financial regulations.
- Different countries have **conflicting laws** on tokenized investments, complicating international film financing.

Mitigation Strategies:

- **Seek legal consultation early** – Filmmakers and GFT issuers should work with **blockchain-savvy entertainment attorneys** to ensure compliance with security laws.
- **Use regulated blockchain platforms** – Projects should consider **SEC-compliant token platforms** (such as

security token offerings [STOs] instead of unregulated NFTs).

👌 **Adopt transparent tokenomics** – Clearly define **investor rights, revenue-sharing models, and payout structures** to avoid legal ambiguity.

Future Outlook:

As **regulatory frameworks mature**, GFT projects will need **structured compliance strategies**—similar to how the film industry evolved with crowdfunding regulations (like the JOBS Act).

2. Market Volatility - Protecting Against Token Devaluation

The Risk:

- The cryptocurrency market **is highly volatile**, meaning GFT values can **fluctuate unpredictably**.
- Investors may **cash out too soon**, destabilizing the film's financing structure.
- **Speculative trading** could overshadow the actual creative value of a film, leading to financial bubbles.

Mitigation Strategies:

👌 **Tie GFTs to real-world utility** – Ensure tokens **offer tangible benefits** (e.g., access to screenings, voting rights) rather than being purely speculative.

👌 **Implement vesting periods** – Require early investors and token holders to hold tokens for a set time to prevent mass sell-offs.

- 👌 **Introduce stablecoin-backed models** – Pegging part of the investment to stable cryptocurrencies (e.g., USDC, DAI) can protect against extreme volatility.

Future Outlook:

More hybrid funding models (blending traditional financing with GFTs) will emerge to reduce dependency on volatile crypto markets.

3. Investor Fraud & Scams - Preventing Rug Pulls and Exploitation

The Risk:

- ⊙ **Bad actors** can create **fraudulent film projects**, raise money via GFTs, and then disappear (**rug pull scams**).
- ⊙ Lack of **oversight and regulation** makes it easier for **unvetted projects to deceive investors**.
- ⊙ **Ponzi-like tokenomics** could emerge, where early investors profit at the expense of later investors.

Mitigation Strategies:

- 👌 **Demand full transparency** – Filmmakers should **publicly disclose project details**, budgets, and distribution plans before launching a GFT sale.
- 👌 **Use escrow smart contracts** – Funds should be **locked in a smart contract** that releases payments **only when key milestones are met** (e.g., production start, completion, release).

- 👌 **Community due diligence** – Encourage **DAOs and investment groups** to conduct thorough **vetting before funding projects.**

Future Outlook:

Blockchain film financing will likely develop **trusted platforms and ratings systems** (similar to **IMDB or Rotten Tomatoes for Web3 films**) to **verify credible projects.**

4. Lack of Web3 Education - Ensuring Widespread Adoption

The Risk:

- Many filmmakers, producers, and investors **lack technical knowledge of blockchain.**
- Misinformation and **scams create distrust** in Web3 financing models.
- The **film industry is slow to adapt to technological change**, making it hard for GFTs to gain mainstream acceptance.

Mitigation Strategies:

- 👌 **Launch Web3 education programs** – Studios, film schools, and industry organizations should **train filmmakers on blockchain financing and smart contracts.**
- 👌 **Create user-friendly platforms** – **Simple, intuitive interfaces** will make GFTs accessible to non-technical creators and investors.

👌 **Partner with traditional film institutions** – Bringing in **Hollywood veterans and established production houses** will **increase trust** in Web3 financing.

Future Outlook:

Web3 adoption in entertainment will grow **exponentially** once **educational resources and user-friendly blockchain tools** become standard in the industry.

5. Distribution Challenges - Ensuring Tokenized Films Reach Audiences

The Risk:

- **GFT-backed films** may struggle to secure **mainstream distribution** without studio partnerships.
- Streaming giants (Netflix, Disney+, HBO) **may resist blockchain-based revenue models** that disrupt their subscription models.
- **Limited Web3-native distribution platforms** make it harder for films to reach large audiences.

Mitigation Strategies:

- 👌 **Integrate with existing distribution channels** – Work with theatrical distributors, streaming platforms, and VOD services that accept blockchain-funded projects.
- 👌 **Create Web3-native film platforms** – Decentralized streaming services (such as **Livepeer or Theta Network**) offer **direct-to-audience monetization** without intermediaries.
- 👌 **Tokenized marketing incentives** – Offer **NFT-based rewards, digital collectibles, or exclusive access** to increase viewer engagement.

Future Outlook:

Major studios and streaming services will eventually adopt hybrid blockchain models to integrate **GFT-powered financing** while maintaining their dominance in distribution.

6. Sustainability Concerns - Addressing the Environmental Impact of Blockchain

The Risk:

- Traditional **blockchain networks (like Ethereum) have high energy consumption**, raising concerns about sustainability.
- Filmmakers and studios may **face backlash** for using energy-intensive blockchain models.

Mitigation Strategies:

- 👌 Use eco-friendly blockchain networks – Transition to low-energy, proof-of-stake (PoS) networks like Ethereum 2.0, Solana, and Polygon.
- 👌 Offset carbon footprints – Invest in carbon-neutral initiatives or partner with sustainability-focused blockchain projects.
- 👌 Educate stakeholders on sustainable blockchain solutions – The industry should promote environmentally responsible Web3 adoption.

Future Outlook:

Blockchain will continue **shifting to eco-friendly models**, reducing concerns over sustainability in film financing.

A Smarter, Safer Web3 Film Economy

While GFTs and blockchain finance offer game-changing opportunities, their long-term success depends on responsible implementation.

Key Takeaways for the Future of GFTs:

1. Regulations will shape the next phase of tokenized film finance. Filmmakers must comply with evolving security laws to ensure legitimacy.
2. Hybrid funding models will emerge. Combining Web3 financing with traditional investment structures will create a balanced, sustainable system.
3. Education and accessibility are crucial. Without proper education, GFTs risk being an exclusive tool for early adopters rather than an industry-wide innovation.
4. Mainstream adoption is inevitable. As major studios, streaming platforms, and DAOs embrace GFTs, decentralized film finance will become the new standard.
5. The shift from Hollywood's outdated studio model to a creator-driven Web3 economy is already underway. The filmmakers, investors, and visionaries who adapt now will define the next golden age of cinema—one built on transparency, equity, and innovation.

REAL-WORLD GFT SUCCESS STORIES

CHAPTER 10

Real-world examples are invaluable for understanding the potential of Web3 fundraising. They illustrate how filmmakers have successfully leveraged GFT innovative tools like NFTs, DAOs, and decentralized platforms to bring their projects to life.

In this chapter, we delve into case studies and inspiring success stories that highlight the practical applications of these strategies, providing actionable insights and demonstrating the transformative power of Film3 in reshaping the film industry.

Case Study #1: "Braid" - Pioneering Tokenized Film Financing

Overview: Braid is a psychological horror film that made history by being one of the first feature films financed through tokenization. In 2018, the filmmakers raised $1.4 million by selling Ethereum-based tokens, allowing investors to own a stake in the film's profits.

- ZONIQX.COM

How GFTs Played a Role:

Democratized Investment: By tokenizing the film's equity, Braid opened investment opportunities to a global audience, enabling supporters to directly fund the project.

Transparent Profit Sharing: Smart contracts ensured that revenue distribution to token holders was automated and transparent, building trust among investors.

Why It Matters:

Innovative Funding Model: Braid demonstrated the potential of blockchain technology to revolutionize film financing, providing a viable alternative to traditional funding avenues.

Empowering Independent Filmmakers: The success of Braid showcased how independent creators could leverage GFTs to maintain creative control while securing necessary funds.

Case Study #2: "Zero Contact" - Embracing NFTs for Film Distribution

Overview: Directed by Rick Dugdale, Zero Contact is a science fiction thriller that embraced NFTs for its distribution strategy. In 2021, the film was released as an NFT, allowing buyers to own a unique copy of the movie and access exclusive content.

- TOKENFI.MEDIUM.COM

How GFTs Played a Role:

Exclusive Ownership: The NFT release provided purchasers with a sense of exclusivity and ownership, differentiating it from traditional digital downloads.

Enhanced Fan Engagement: NFT holders were granted access to special features, behind-the-scenes content, and interactive experiences, fostering a deeper connection with the audience.

Why It Matters:

Redefining Distribution Channels: Zero Contact showcased how NFTs could be utilized to distribute films directly to consumers, bypassing traditional platforms and intermediaries.

Monetizing Digital Assets: The film's approach highlighted new revenue streams for filmmakers through the sale of digital collectibles and exclusive content.

Case Study #3: "Stoner Cats" - Animated Series Funded by NFTs

Overview: Stoner Cats is an animated web series that utilized NFTs to fund its production. In 2021, the creators sold unique NFTs that granted buyers access to watch the series, effectively raising $8 million in under 35 minutes.

- TOKENFI.MEDIUM.COM

How GFTs Played a Role:

Access Tokens: Purchasing an NFT not only supported the project's funding but also served as a pass to view the content, integrating ownership with access.

Community Building: NFT holders became part of an exclusive community with potential future benefits, such as input on content direction and early access to new releases.

Why It Matters:

Innovative Content Monetization: Stoner Cats exemplified how creators could monetize content upfront, reducing financial risks associated with traditional production models.

Engaging Niche Audiences: The project demonstrated the effectiveness of targeting specific audiences willing to invest in unique content experiences.

These case studies exemplify the transformative potential of GFTs and decentralized funding in the film industry. By embracing blockchain technology, filmmakers can explore innovative financing and distribution models, fostering direct engagement with audiences and redefining traditional industry paradigms.

Success Stories: Transforming Film Fundraising with Web3/Film3 Strategies

The rise of Web3 and Film3 has introduced groundbreaking approaches to film financing, using blockchain technology, NFTs, and decentralized platforms to empower creators and engage audiences. By bypassing traditional industry gatekeepers, these strategies open doors for innovation and inclusivity. Below, we explore success stories, including *Julie's Dream* and Film DAO, alongside other noteworthy examples, to demonstrate the practical impact and offer actionable insights.

1. Julie's Dream: Empowering Underserved Voices

The Strategy:

Julie's Dream, a Film3 project, focused on amplifying underrepresented voices in the film industry by leveraging NFTs and community-driven fundraising. The film's creators issued a series of NFTs offering unique benefits, such as direct access to creative updates, premiere tickets, and exclusive digital art. The project also formed partnerships with blockchain organizations committed to diversity and inclusion.

Results:

- Successfully raised over $500,000 through NFT sales.
- Created a dedicated community of supporters who championed the project's mission.
- Drew significant attention as a model for socially driven Film3 initiatives.

Actionable Insights:

- **Highlight a Mission:** Align NFT campaigns with a clear purpose, appealing to both financial and emotional investment.
- **Forge Strategic Partnerships:** Collaborate with blockchain communities to amplify reach and credibility.

2. Film DAO: Community Governance in Action

The Strategy:

Film DAO, a decentralized autonomous organization, was established to democratize film financing by pooling resources from its members. Contributors purchased governance tokens, allowing them to vote on which projects the DAO would fund. Filmmakers submitted proposals to the DAO, presenting their creative visions and funding requirements.

Results:

- Funded several independent films, showcasing a collaborative approach to decision-making.
- Empowered smaller creators who lacked traditional funding opportunities.
- Created a replicable framework for other creative DAOs, sparking a trend in decentralized filmmaking.

Actionable Insights:

- **Promote Transparency:** Use smart contracts to clearly outline funding terms and ensure trust among contributors.
- **Engage in Co-Creation:** Invite token holders to participate in creative decision-making, fostering a sense of ownership.

3. Kevin Smith's "KillRoy Was Here": NFTs as Distribution Models

The Strategy:

Kevin Smith released *KillRoy Was Here* as an NFT, making it one of the first films distributed entirely on the blockchain. Fans who purchased the NFT gained exclusive ownership rights, behind-the-scenes access, and opportunities to participate in future projects.

Results:

- Raised significant funds from NFT sales while generating media buzz.
- Demonstrated a successful alternative to traditional distribution channels.
- Solidified Smith's position as a pioneer in the Film3 space.

Actionable Insights:

- **Monetize Distribution:** Use NFTs not just for funding but also for alternative distribution models.
- **Offer Unique Perks:** Combine NFTs with exclusive experiences to enhance their appeal.

4. "The Infinite Machine": Decentralized Storytelling

The Strategy:

Director Camila Russo used a DAO and Ethereum-based NFTs to fund *The Infinite Machine*, a film about Ethereum's origins. The DAO model allowed contributors to fund the film while participating in its creative journey.

Results:

- Raised over $1.5 million via NFT sales.
- Established a collaborative, community-driven production process.
- Attracted a global audience invested in Ethereum's story and blockchain innovation.

Actionable Insights:

- **Leverage Community Passion:** Tap into niche communities aligned with your film's themes for organic support.
- **Incorporate Decentralization:** Explore DAOs for collaborative decision-making and governance.

5. "Stoner Cats": NFT Crowdfunding Success

The Strategy:

Stoner Cats, an animated series featuring Mila Kunis, funded its production through the sale of exclusive NFTs. NFT holders gained access to episodes and community perks, creating a direct relationship with the audience.

Results:

- Raised $8 million in under 40 minutes by selling out its NFT collection.
- Fostered a tight-knit community of backers who acted as brand ambassadors.
- Demonstrated the viability of NFTs as a direct-to-consumer funding strategy.

Actionable Insights:

- **Create Urgency:** Use limited-edition NFT drops to drive demand.
- **Bundle Benefits:** Enhance NFT value with exclusive content and experiences.

6. "Julie's Dream" and Film DAO: The Future of Inclusive Storytelling

These two projects exemplify how Film3 strategies can address industry inequities and empower creators:

- *Julie's Dream* highlights how NFTs can amplify diverse voices and engage audiences around a cause.
- Film DAO showcases how decentralized governance can fund independent projects democratically, creating opportunities for creators outside traditional power structures.

By integrating blockchain tools, these projects offer a blueprint for sustainable, community-driven filmmaking.

Takeaways for Filmmakers:

1. **Align with a Mission:** Projects like *Julie's Dream* prove that a strong purpose can drive funding and community support.
2. **Leverage Blockchain Transparency:** Platforms like Film DAO highlight the importance of using smart contracts for accountability and trust.
3. **Offer Long-Term Value:** Successful campaigns ensure NFT or token holders receive ongoing perks, such as exclusive content or profit-sharing opportunities.
4. **Educate Your Audience:** Simplify the onboarding process for backers new to Web3 to expand your reach.

Web3 and Film3 are reshaping the film industry by introducing inclusive, transparent, and community-driven models for funding and storytelling. Success stories like *Julie's Dream*, Film DAO, and others demonstrate the transformative potential of these strategies, empowering filmmakers to innovate while creating deeper connections with their audiences. By embracing these insights, creators can tap into the future of filmmaking and bring their visions to life.

NAVIGATING CHALLENGES IN WEB3 FUNDRAISING

CHAPTER 11

Web3 fundraising offers exciting opportunities, but it also comes with unique challenges that filmmakers must navigate. From understanding complex legal and regulatory landscapes to educating a diverse audience on blockchain basics, these hurdles require preparation and adaptability. Additionally, managing risks like scams and market volatility is crucial to ensuring the success and security of your fundraising campaign.

This chapter explores the most common challenges in Web3 fundraising and provides practical strategies to overcome them.

Web3, the decentralized internet powered by blockchain technology, has introduced groundbreaking opportunities in fundraising. From Initial Coin Offerings (ICOs) to Security Token Offerings (STOs) and Decentralized Autonomous Organization (DAO)-based funding, the Web3 space has transformed how startups and projects secure capital. However, this frontier also presents unique challenges that require tailored strategies. Below, we delve into the most common hurdles in Web3 fundraising and outline practical approaches to navigate them.

1. Regulatory Uncertainty

The Challenge:

Regulations around blockchain based fundraising vary widely across jurisdictions and remain a gray area in many regions. Projects often struggle to ensure compliance while still attracting global investors. Non-compliance can lead to hefty fines, project shutdowns, or loss of investor trust.

Strategies to Overcome:

- **Engage Legal Experts:** Work with legal advisors experienced in blockchain to navigate local and international laws.
- **Adopt a Compliance-First Approach:** Opt for fundraising models like STOs that are designed to adhere to securities laws.
- **Stay Updated:** Continuously monitor regulatory developments to adapt your approach proactively.
- **Implement KYC/AML Protocols:** Use robust Know Your Customer (KYC) and Anti-Money Laundering (AML) solutions to ensure regulatory adherence.

2. Market Volatility

The Challenge:

Cryptocurrency markets are notoriously volatile, which can affect the valuation of funds raised and lead to investor hesitation.

Strategies to Overcome:

- **Stablecoins for Fundraising:** Accept contributions in stablecoins like USDC or DAI to mitigate valuation risks.
- **Diversified Fund Management:** Convert a portion of raised funds into fiat or stable assets to hedge against volatility.

- **Transparent Communication:** Maintain open dialogue with investors about how market conditions are being managed.

3. Lack of Trust and Awareness

The Challenge:

Many potential investors are unfamiliar with Web3 technologies or wary of scams, which are prevalent in the space.

Strategies to Overcome:

- **Educational Outreach:** Create comprehensive resources, webinars, and tutorials to educate potential investors about your project and Web3 basics.
- **Third-Party Audits:** Conduct smart contract audits and security checks by reputable firms to build trust.
- **Transparent Roadmap and Team:** Publish a clear roadmap and detailed team profiles to demonstrate credibility and accountability.

4. High Competition

The Challenge:

The Web3 ecosystem is crowded with numerous projects vying for attention and capital. Standing out is a significant challenge.

Strategies to Overcome:

- **Unique Value Proposition:** Clearly articulate how your project solves a problem or addresses a gap in the market.

- **Community Engagement:** Build an active community through platforms like Discord, Twitter, and Telegram.
- **Leverage Influencers and Partnerships:** Collaborate with thought leaders and other projects to amplify visibility.
- **Consistent Branding:** Develop a cohesive and memorable brand identity that resonates with your target audience.

5. Technical Barriers

The Challenge:

Investors unfamiliar with blockchain technology often face challenges in participating due to technical requirements like wallet setup or token acquisition.

Strategies to Overcome:

- **User-Friendly Interfaces:** Simplify the fundraising process with intuitive platforms and guides.
- **Step-by-Step Tutorials:** Offer clear, beginner-friendly instructions for setting up wallets and contributing.
- **Support Channels:** Provide accessible support via live chat or FAQs to assist investors in real time.

6. Tokenomics Mismanagement

The Challenge:

Poorly designed tokenomics can lead to inflation, misaligned incentives, or an unsustainable economic model, deterring investors.

Strategies to Overcome:

- **Expert Consultation:** Collaborate with tokenomics specialists to design a sustainable model.
- **Incentive Alignment:** Structure tokens to reward long-term engagement rather than speculative behavior.
- **Regular Reviews:** Continuously analyze and adjust tokenomics based on market conditions and community feedback.

7. Security Concerns

The Challenge:

Web3 fundraising platforms and smart contracts are prime targets for cyberattacks, which can erode investor confidence and result in significant financial losses.

Strategies to Overcome:

- **Audits and Penetration Testing:** Regularly audit smart contracts and conduct penetration tests to identify vulnerabilities.
- **Multi-Signature Wallets:** Use multi-signature wallets to secure funds and prevent unauthorized access.
- **Bug Bounty Programs:** Encourage ethical hackers to identify and report potential weaknesses in exchange for rewards.

While Web3 fundraising offers unparalleled opportunities, navigating its challenges requires strategic planning, technical expertise, and a commitment to transparency. By addressing regulatory uncertainty, market volatility, trust deficits, and other common obstacles, projects can position themselves for successful fundraising in this transformative ecosystem.

As the Web3 landscape evolves, staying adaptable and proactive will be key to thriving in this dynamic space.

EXPANDING YOUR REACH – ENTER THE METAVERSE

CHAPTER 12

The metaverse is revolutionizing the way stories are told and consumed, offering filmmakers an entirely new medium for creativity and audience engagement. By blending digital and physical realities, the metaverse allows creators to break free from traditional constraints, opening doors to immersive storytelling, interactive premieres, and innovative distribution methods.

This chapter explores how filmmakers can harness the potential of the metaverse to expand their reach, deepen audience connections, and redefine the future of cinema.

Exploring the Metaverse: Redefining Cinema for Filmmakers

The metaverse, a digital realm where virtual and augmented realities converge, is no longer just a concept of science fiction—it is rapidly becoming a cornerstone of modern interaction, commerce, and creativity. For filmmakers, the metaverse offers unprecedented opportunities to expand their reach, deepen connections with audiences, and reshape the boundaries of storytelling. Here's how filmmakers can harness its potential to redefine the future of cinema.

1. Expanding Audience Reach

Global Accessibility

The metaverse eliminates geographical barriers, making films accessible to audiences worldwide in immersive, shared environments. Virtual reality (VR) theaters or metaverse-based film festivals allow global audiences to gather and experience films without physical constraints.

New Distribution Channels

Platforms like Decentraland, Spatial, and VRChat are emerging as hubs for creative content. Filmmakers can release their work in these digital spaces, bypassing traditional gatekeepers and reaching younger, tech-savvy demographics who are already active in the metaverse.

Microeconomies and NFTs

In the metaverse, filmmakers can monetize their work in innovative ways, such as selling Non-Fungible Tokens (NFTs). NFTs could include limited-edition digital posters, exclusive behind-the-scenes footage, or even unique film frames that fans can own. These provide new revenue streams while creating a sense of exclusivity and engagement.

2. Deepening Audience Connections

Interactive Storytelling

The metaverse transforms audiences from passive viewers into active participants. Imagine a VR film where viewers can influence the storyline, explore scenes from different perspectives, or interact with characters. This level of interactivity fosters deeper emotional engagement and loyalty.

Virtual Communities

Filmmakers can build vibrant fan communities in the metaverse. Virtual premieres, Q&A sessions with directors, or workshops can take place in 3D environments, giving audiences a sense of personal connection and investment in the creative process.

Personalized Experiences

With AI and user data, filmmakers can craft customized experiences in the metaverse. Personalized narratives or VR environments tailored to individual users could revolutionize how stories resonate with different viewers.

3. Redefining the Future of Storytelling

Immersive World-Building

The metaverse allows filmmakers to extend their narratives into expansive, explorable worlds. Fans can step into a fully realized digital universe, blurring the line between cinema and gaming. A sci-fi epic, for example, could have a metaverse-based prequel experience that lets users discover the backstory interactively.

Collaborative Creativity

In the metaverse, filmmakers can collaborate with digital artists, animators, and even audiences to co-create content. Open-source storylines or crowd-funded projects could emerge, giving rise to a more democratic filmmaking process.

Virtual Production

The metaverse is poised to revolutionize how films are made. Virtual production tools, such as Epic Games' Unreal Engine, are already being used to create photorealistic sets and environments. These tools reduce costs and allow for real-time visualization of complex scenes.

4. Challenges and Considerations

Technical Expertise

Filmmakers need to familiarize themselves with metaverse technologies, such as VR, AR, blockchain, and AI. Partnering with tech-savvy creators and developers can ease this transition.

Cost and Accessibility

While the metaverse offers cost-saving opportunities in some areas, initial investments in VR equipment, software, and expertise may be prohibitive. Ensuring accessibility for diverse audiences is also crucial to achieving widespread adoption.

Ethical Implications

Filmmakers must navigate ethical concerns, including privacy, data security, and inclusivity, as the metaverse evolves. Responsible storytelling and platform management will be essential to building trust with audiences.

5. Practical Steps for Filmmakers to Get Started

1. **Experiment with Short Films**: Start by creating short VR or AR experiences to familiarize yourself with metaverse storytelling.
2. **Partner with Tech Experts**: Collaborate with developers and metaverse platforms to explore distribution and interaction options.
3. **Engage Fans Early**: Involve your audience in the creative process through metaverse platforms, fostering a sense of ownership and anticipation.
4. **Leverage NFTs and Blockchain**: Use these tools for funding, merchandising, and creating unique experiences that reward loyal fans.

The metaverse is a dynamic, evolving landscape that offers limitless potential for filmmakers willing to embrace its possibilities. By leveraging its capabilities for global distribution, interactive storytelling, and immersive world-building, filmmakers can redefine the cinematic experience for a new generation. As technology and creativity continue to intersect, the metaverse may well become the ultimate stage for the future of cinema.

THE METAVERSE: A NEW FRONTIER FOR FILM DISTRIBUTION AND IMMERSIVE EXPERIENCES

CHAPTER 13

Monetizing a Film in the Metaverse

The metaverse, a collective virtual shared space, represents a groundbreaking frontier for the film industry. This digital realm, blending augmented reality (AR), virtual reality (VR), and blockchain technology, offers creators and producers innovative ways to monetize their films beyond traditional theaters or streaming platforms.

In this chapter we provide is a detailed exploration of the diverse monetization strategies available in the metaverse.

1. NFT-Based Monetization

Non-fungible tokens (NFTs) are digital assets that provide unique ownership rights to virtual goods. For filmmakers, NFTs unlock opportunities for direct fan engagement and revenue generation.

- **Collectible NFTs**: Sell film-related digital collectibles, such as concept art, character designs, or exclusive behind-the-scenes footage. These tokens can serve as memorabilia for fans and generate buzz for the film.

- **Exclusive Content**: Offer special editions of scenes, director's cuts, or alternate endings as NFTs. Ownership of these assets can confer bragging rights and a sense of exclusivity.

- **Royalties Through Smart Contracts**: Smart contracts associated with NFTs enable creators to earn a percentage of resale profits when the NFTs change hands, creating a recurring revenue stream.

2. Virtual Screenings and Ticket Sales

The metaverse allows filmmakers to host virtual premieres and screenings within immersive environments.

- **Paid Virtual Screenings**: Charge entry fees for audiences to attend VR-based premieres or viewings. This approach replicates the excitement of red-carpet events within a digital framework.
- **Subscription-Based Access**: Provide access to a series of films or exclusive content through a subscription model.
- **Dynamic Viewing Experiences**: Leverage the interactive capabilities of the metaverse by offering viewers the ability to choose camera angles, follow specific characters, or even influence story outcomes.

3. Virtual Merchandise and In-World Sales

In the metaverse, the concept of merchandise transcends physical goods. Virtual merchandise can enhance the film's brand presence and provide fans with interactive experiences.

- **Character Avatars and Costumes**: Sell digital versions of character outfits or avatars that users can wear in metaverse environments.
- **Virtual Props**: Offer props from the film as 3D assets for metaverse users to incorporate into their digital spaces.
- **Exclusive In-World Purchases**: Create themed environments or spaces where users can buy limited-edition items directly related to the film's storyline.

4. Metaverse-Based Advertising and Partnerships

Films in the metaverse can serve as platforms for branded partnerships and advertisements, opening new revenue streams.

- **Product Placement in Virtual Worlds**: Integrate branded products into the film's metaverse environment, ensuring subtle yet effective exposure.
- **Branded Virtual Events**: Partner with brands to host events, such as interactive meet-and-greets or live Q&A sessions with the cast, within the film's virtual space.
- **Sponsorship Opportunities**: Collaborate with tech or lifestyle brands to sponsor film-related experiences, such as exclusive previews or gamified interactions.

5. Gamification and Interactive Storytelling

Gamification involves transforming elements of the film into engaging, interactive experiences.

- **Story-Based Games**: Develop games inspired by the film's plot, where users can take on roles within the narrative. Offer premium access or in-game purchases as monetization methods.
- **Interactive Storylines**: Allow users to influence the story through their decisions, creating personalized viewing experiences and monetizing access to unique story arcs.
- **Achievements and Rewards**: Encourage users to complete challenges related to the film to unlock exclusive rewards, which can be monetized as NFTs or digital goods.

6. Community Building and Fan Engagement

Building a loyal fan base in the metaverse can lead to sustained revenue opportunities.

- **Membership Programs**: Offer tiered memberships providing access to perks such as exclusive content, virtual meetups with cast and crew, and early access to new releases.

- **Crowdfunding Campaigns**: Use the metaverse to host interactive crowdfunding campaigns, offering backers unique rewards such as NFT-based credits or early screenings.
- **Virtual Film Festivals**: Organize virtual festivals where multiple filmmakers showcase their work, generating income through ticket sales or sponsorship deals.

7. Blockchain-Enabled Revenue Sharing

Blockchain technology provides transparency and efficiency in revenue distribution.

- **Decentralized Distribution Platforms**: Leverage blockchain to distribute films directly to consumers, bypassing traditional intermediaries and retaining a greater share of profits.
- **Tokenized Revenue Sharing**: Issue tokens representing shares in the film's revenue, allowing fans and investors to participate in its financial success.

8. Integration with Existing Metaverse Platforms

Collaboration with established metaverse platforms can amplify reach and monetization potential.

- **Partnerships with Metaverse Giants**: Platforms like Decentraland, The Sandbox, or Meta's Horizon Worlds offer pre-established user bases and infrastructure for hosting virtual events and selling digital goods.
- **Cross-Promotion with Other Creators**: Collaborate with creators within the metaverse to cross-promote films

and shared experiences, driving mutual growth and revenue.

The metaverse provides an expansive playground for filmmakers to experiment with creative monetization strategies. From NFTs and gamified storytelling to branded partnerships and virtual events, the opportunities are as limitless as the imagination of the creators. By embracing this digital frontier, filmmakers can not only unlock new revenue streams but also foster deeper connections with audiences in innovative and engaging ways. The future of film monetization lies in the metaverse—and those who adapt early will undoubtedly lead the charge.

NFT MOVIES AS AN ADD-ON TO TRADITIONAL DISTRIBUTION

CHAPTER 14

As filmmakers explore new avenues for reaching audiences, NFTs have emerged as a powerful complement to traditional distribution models. By offering unique digital assets, filmmakers can enhance fan engagement, generate additional revenue, and expand their project's reach.

This chapter delves into how NFTs can be integrated into traditional film distribution, providing real-world examples and actionable strategies to bridge the gap between conventional and cutting-edge approaches.

What Are NFT Movies?

NFT movies leverage blockchain technology to offer digital collectibles and exclusive content tied to a film. These NFTs can complement traditional distribution by creating unique, value-added experiences for fans and collectors.

Integrating NFTs into Traditional Film Distribution

The advent of Non-Fungible Tokens (NFTs) has transformed industries ranging from art to gaming, presenting an unprecedented opportunity for the film industry. By integrating NFTs into traditional film distribution, studios and creators can unlock new revenue streams, enhance audience engagement, and create innovative marketing strategies. This chapter explores how NFTs can bridge the gap between conventional film distribution methods and cutting-edge technologies, offering actionable strategies and real-world examples for higher returns on investment (ROI).

The Potential of NFTs in Film Distribution

NFTs, digital assets verified on a blockchain, are unique and non-interchangeable. In the context of film, NFTs can represent collectibles, exclusive content, access rights, or even investment stakes. The decentralization and traceability provided by blockchain technology ensure authenticity and transparency, making NFTs an appealing proposition for both creators and audiences.

Value Proposition of NFTs for the Film Industry:

- **Enhanced Fan Engagement:** Fans can own a piece of the film's history, such as concept art, scripts, or exclusive behind-the-scenes footage.
- **Alternative Revenue Streams:** Studios can monetize digital assets, reducing dependence on traditional box office and streaming revenues.
- **Promotion and Marketing:** Unique digital assets can generate buzz and provide early funding opportunities.
- **Audience Expansion:** NFTs can attract crypto-enthusiasts, expanding the demographic reach of films.

Real-World Examples

1. "Zero Contact" NFT Release

In 2021, the film *Zero Contact* starring Anthony Hopkins became one of the first feature films released as an NFT. Vuele, the platform behind the NFT release, allowed fans to own exclusive versions of the film, special content, and memorabilia. This approach not only garnered significant media attention but also generated a direct revenue stream from a dedicated audience.

2. Quentin Tarantino's Pulp Fiction NFTs

Quentin Tarantino released NFTs of uncut scenes and original handwritten scripts from *Pulp Fiction*. These NFTs were auctioned off, providing fans and collectors with rare access to previously unseen elements of the iconic film.

3. Kevin Smith's "Killroy Was Here"

Kevin Smith's horror-comedy film was released as an NFT, allowing the buyer to distribute it traditionally or keep it exclusive. This novel approach highlighted how NFTs could serve as both a distribution and investment tool.

Strategies to Bridge Traditional and NFT Distribution

1. Tokenized Ownership for Crowdfunding

Film studios can offer tokenized ownership stakes via NFTs during pre-production. Fans who invest through NFT purchases could receive a share of future profits or exclusive content. For example:

- An indie film could issue NFTs representing 1% ownership each, with buyers receiving proportional revenue from box office and streaming earnings.

2. Exclusive Digital Collectibles

NFTs can represent limited-edition digital posters, signed scripts, or virtual meet-and-greet passes with cast members. These collectibles can be bundled with tickets or streaming subscriptions to enhance value.

3. Enhanced Streaming Experiences

Streaming platforms can integrate NFT-based perks, such as access to directors' cuts, alternate endings, or interactive content. This approach aligns NFTs with existing subscription models, adding layers of exclusivity.

4. Bridging Physical and Digital Worlds

NFTs can combine digital assets with physical experiences. For example:

- Purchase an NFT of a behind-the-scenes clip and also receive a VIP ticket to the film's premiere.
- Own an NFT of a film scene and get a physical, autographed storyboard.

5. Gamification and Loyalty Programs

NFTs can gamify the film experience, rewarding loyal fans with tokens redeemable for merchandise, access to premieres, or credits in future productions. Blockchain technology ensures transparent tracking of fan engagement.

Overcoming Challenges

1. Accessibility and Education

Many traditional film audiences are unfamiliar with blockchain and NFTs. Studios need to invest in user-friendly platforms and educational campaigns to bridge the knowledge gap.

2. Legal and Regulatory Issues

Tokenized ownership and intellectual property rights can be complex. Collaborating with legal experts to navigate regulations is crucial.

3. Sustainability Concerns

The environmental impact of blockchain technology, particularly energy-intensive networks, remains a concern. Adopting eco-friendly blockchain solutions, such as Ethereum's shift to Proof of Stake, can mitigate these issues.

Integrating NFTs into traditional film distribution offers a symbiotic blend of innovation and nostalgia. By leveraging blockchain technology, filmmakers can create unique value propositions for their audiences, optimize marketing strategies, and diversify revenue streams. As the case studies and strategies in this chapter illustrate, the potential for NFTs in film is vast and largely untapped.

By embracing this emerging technology with actionable strategies, studios and independent creators alike can navigate the evolving landscape of film distribution, ensuring relevance and profitability in an increasingly digital world.

INCORPORATING MARKETING, PR, AND KEY ART FOR SUCCESSFUL OUTCOMES IN WEB3/FILM3

CHAPTER 15

The emergence of Web3 and Film3 has redefined the way creators, brands, and audiences engage in storytelling and content distribution. These decentralized platforms leverage blockchain, NFTs, and smart contracts to disrupt traditional film and media industries. However, success in Web3/Film3 relies not just on innovative technology but also on strategic marketing, public relations (PR), and compelling key art. Here's how these components can be integrated effectively:

1. Marketing in Web3/Film3: Building Awareness and Community

Web3 and Film3 projects thrive on community engagement, transparency, and innovative value propositions. Marketing in this space requires a blend of traditional strategies and Web3-specific tactics:

a. Narrative Development

- **Importance**: A strong narrative is essential to connect with audiences who value authenticity and purpose.
- **Implementation**: Position the project within the ethos of decentralization and creator empowerment. Highlight unique aspects, such as blockchain technology, NFT utility, or community co-ownership.
- **Example**: Projects like *The Ethereum Movie Venture* used tokenized storytelling to resonate with Web3 ethos.

b. Multi-Channel Strategy

- **Web3 Platforms**: Engage audiences on Discord, Twitter, Telegram, and decentralized social networks (e.g., Lens Protocol).
- **Traditional Media**: Use legacy platforms like YouTube, Instagram, and TikTok for broader reach.
- **Content**: Create teaser trailers, behind-the-scenes insights, AMAs (Ask Me Anything), and explainer videos about tokenomics and NFTs.

c. Data-Driven Personalization

Leverage analytics tools to understand audience behavior on blockchain platforms. Tailor experiences to their preferences, whether by exclusive NFT drops, access to dApps, or virtual events in the metaverse.

2. Public Relations (PR): Building Credibility and Trust

In Web3/Film3, PR extends beyond press releases to establish a project's legitimacy and create excitement:

a. Credibility through Strategic Partnerships

- **Why It Matters**: The decentralized nature of Web3 can be intimidating to newcomers. Partnerships with established creators, brands, and platforms lend credibility.
- **Execution**: Collaborate with influential figures in the blockchain or entertainment industries and feature them prominently in campaigns.

- **Example**: The Bored Ape Yacht Club partnered with film studios to enhance their storytelling, creating an aspirational aura around their brand.

b. Media Outreach

Develop tailored pitches for Web3-focused outlets like *Decrypt*, *CoinDesk*, and *NFT Now*, while also engaging mainstream media like *The Hollywood Reporter*.

- Showcase innovations such as fractionalized ownership models or DAO-driven production.
- Ensure journalists receive easy-to-understand explanations of the project's value.

c. Community Relations

- **Transparency**: Regularly communicate updates about milestones, challenges, and roadmap changes. Platforms like Medium and Mirror.xyz are excellent for long-form updates.
- **Interactive Engagement**: Host live Q&A sessions or town halls on Twitter Spaces and Discord to involve the community in decision-making.

3. Key Art: Visual Storytelling to Captivate and Convert

Key art serves as the visual representation of your project, making it a critical component in a space driven by NFTs and digital art.

a. Iconic NFT Design

- **Significance**: NFTs are both collectibles and marketing tools. Their design should encapsulate the essence of your story while providing functional utility.
- **Strategies**:
    - Collaborate with prominent digital artists in the NFT space to create iconic imagery.
    - Utilize generative art to add rarity and uniqueness.
    - Example: *Beeple's EVERYDAYS* not only achieved visual appeal but also became a cultural symbol.

b. Branding Consistency

- Develop a cohesive aesthetic for posters, promotional banners, and social media graphics that align with the project's narrative and technical innovation.
- Incorporate Web3 elements like blockchain symbols, QR codes for NFTs, or augmented reality (AR) features.

c. Interactive Visuals

- Create AR-enhanced posters or key art that allow audiences to scan and view 3D content or trailers. These interactive elements can be tied to NFT ownership, unlocking exclusive features.
- Example: Projects like *Matrix Resurrections NFT* integrated AR into promotional campaigns to amplify audience engagement.

4. Integrating Components for Holistic Success

To achieve the best outcomes, marketing, PR, and key art should function as interconnected elements, creating a seamless and immersive experience.

a. Phased Rollouts

- **Pre-Launch**: Focus on community-building and education. Use teasers, explainer videos, and sneak peeks of NFT collections.
- **Launch Phase**: Leverage key art for maximum visibility on marketplaces like OpenSea and Rarible. Simultaneously, execute PR campaigns targeting Web3 media and traditional outlets.
- **Post-Launch**: Use NFT airdrops, metaverse screenings, or DAO-driven content releases to maintain momentum.

b. Cross-Platform Synergy

- Promote key art in PR materials to create visual consistency.
- Share key marketing updates through NFTs, such as "roadmap reveal posters" or "unlockable content."

In Web3/Film3, the synergy of marketing, PR, and key art is crucial for creating immersive, community-driven projects that resonate with audiences and investors alike. The key lies in maintaining authenticity while leveraging cutting-edge technology and creativity to build trust and engagement. By doing so, creators can redefine storytelling and distribution, aligning with the values of decentralization and empowerment that Web3 represents.

EXPLORING TRADITIONAL FUNDRAISING APPROACHES FOR FILMMAKERS AND THE POTENTIAL OF WEB3 INTEGRATION

CHAPTER 16

While Web3 and Film3 are revolutionizing how filmmakers approach funding, traditional fundraising methods remain highly effective and essential. These methods often provide stable, proven pathways for securing financial resources and reaching key stakeholders in the industry.

Funding a film has always been a daunting yet integral part of the filmmaking process. From personal savings to partnerships with studios, the landscape is filled with opportunities and challenges. This chapter examines traditional fundraising techniques, ranked from the easiest to the most challenging, and explores how innovative Web3 tools can complement these methods to maximize success.

1. Personal Savings: The Starting Point

Difficulty Level: Easy

Overview: Personal savings are often the first resource filmmakers tap into. This approach is straightforward, requiring no external validation or approval. Filmmakers retain full control over their projects, ensuring creative freedom.

Challenges:

- Limited resources: Depending on personal financial capacity, this method can only fund small-scale projects or serve as seed money.
- Risk of personal loss: Investing one's own money can strain personal finances, especially if the project doesn't yield financial returns.

Web3 Synergy:

Personal savings can be augmented by tokenizing part of the project's future revenue on blockchain platforms. This enables filmmakers to raise additional funds without losing creative control, while also offering early supporters a tangible stake in the film's success.

2. Crowdfunding: Engaging the Community

Difficulty Level: Moderate

Overview: Platforms like Kickstarter and Indiegogo have democratized film financing, allowing filmmakers to pitch their ideas directly to audiences. With a compelling campaign, filmmakers can gather small contributions from numerous backers.

Challenges:

- Competition: The crowded nature of crowdfunding platforms requires standout marketing and creative pitches.
- Reliance on marketing: Success often hinges on social media reach and effective storytelling to gain visibility.

Web3 Synergy:

Decentralized crowdfunding platforms take traditional crowdfunding to the next level. Using blockchain, filmmakers can issue NFTs or tokens tied to exclusive rewards, such as digital collectibles or behind-the-scenes access. This incentivizes backers while broadening the appeal of the campaign.

3. Grants: Funding Through Institutions

Difficulty Level: Moderate to Challenging

Overview: Grants from film foundations, arts councils, or government bodies provide filmmakers with funding without the expectation of repayment or relinquishment of rights. These programs often prioritize projects with cultural, social, or artistic value.

Challenges:

- Stringent eligibility criteria: Filmmakers must align their projects with the grantor's objectives, which can limit creative freedom.
- Highly competitive: Applications can be time-consuming and may not always yield results.

Web3 Synergy:

By combining grant applications with Web3 tools, filmmakers can build community support for their proposals. For instance, NFT sales can demonstrate audience interest and provide additional funding leverage, enhancing a project's appeal to grant committees.

4. Private Investors: Leveraging High-Net-Worth Individuals

Difficulty Level: Challenging

Overview: Securing funding from private investors requires persuasive pitches and a solid business plan. These investors often seek a return on investment, making financial viability a critical aspect of the proposal.

Challenges:

- High stakes: Private investors typically demand accountability and may impose creative constraints.
- Building trust: Filmmakers must establish credibility and demonstrate their project's potential for success.

Web3 Synergy:

Smart contracts on blockchain platforms can bring transparency to investor relationships. Filmmakers can outline revenue-sharing agreements on-chain, ensuring investors are paid accurately and promptly, fostering trust.

5. Studio Partnerships: Industry-Endorsed Funding

Difficulty Level: Very Challenging

Overview: Partnerships with major studios provide significant funding and access to vast resources like marketing and distribution networks. Studios often seek projects with mainstream appeal or established talent.

Challenges:

- Creative compromises: Studios may demand changes to ensure commercial viability.
- Gatekeeping: Breaking into the industry and pitching successfully to a studio can be nearly impossible for independent filmmakers without connections or a strong portfolio.

Web3 Synergy:

Filmmakers can use decentralized platforms to showcase their project's traction. By proving their project's popularity or profitability through NFT presales or tokenized ownership, they can demonstrate to studios the value of collaboration while retaining some creative independence.

Combining Traditional Techniques with Web3 Tools

The integration of Web3 tools into traditional fundraising strategies offers filmmakers unparalleled opportunities to bridge the gap between art and finance.

Here's how combining these approaches can enhance success:

1. **Community Empowerment:** Web3 enables direct engagement with audiences through NFTs, tokens, or DAOs (Decentralized Autonomous Organizations). Filmmakers can build grassroots support that complements traditional methods like crowdfunding or investor pitches.
2. **Transparency and Trust:** Blockchain's transparency provides a secure and immutable record of transactions, whether for crowdfunding backers or private investors, ensuring accountability.
3. **New Revenue Streams:** NFTs and tokenized ownership can generate revenue pre- and post-release, helping to offset production costs and creating long-term income streams.

4. **Global Reach:** Web3 tools transcend geographical barriers, allowing filmmakers to access global audiences and investors who might otherwise remain out of reach.

Raising funds for a film is both an art and a science. From leveraging personal savings to securing studio partnerships, each traditional approach comes with its own set of challenges. However, the integration of Web3 tools like NFTs, tokenized ownership, and blockchain-based crowdfunding can revolutionize the way filmmakers approach financing. By combining the best of both worlds, filmmakers can diversify their funding sources, maintain creative independence, and maximize their chances of success in a rapidly evolving industry.

In the era of digital transformation, the fusion of traditional and innovative methods is not just an option—it's the future.

Traditional Fundraising Approaches for Filmmakers from Easy to Difficult

1. **Personal Savings**: Using your own money to finance your film. This approach requires no external validation but may be limited by your personal financial resources.

2. **Friends and Family**: Asking for financial support from close contacts. While easier to access, it requires a strong pitch to gain their trust and investment.

3. **Local Grants and Competitions**: Applying for small-scale grants or local film competitions. These often have clear requirements and accessible entry points.

4. **Crowdfunding Platforms**: Using sites like Kickstarter or Indiegogo. This involves creating a campaign and leveraging your network to reach funding goals.

5. **Private Investors**: Pitching your project to angel investors or venture capitalists who are interested in creative ventures. This can be time-consuming and requires a convincing business plan.
6. **Film Festivals and Markets**: Showcasing your project to attract distributors or funding. This requires a polished presentation and networking skills.
7. **Studio or Production Company Funding**: Partnering with a studio or production house. This involves navigating competitive pitches and potential compromises on creative control.
8. **Bank Loans and Financing**: Taking out loans or accessing traditional financing options. This is often the most difficult due to the need for collateral and the financial risk involved.

WEB3 TERMS TO KNOW FOR FILM3 AND WHY THEY MATTER

CHAPTER 17

Film3 refers to the integration of Web3 technologies into the film industry, enabling creators to rethink traditional models for funding, distribution, and audience engagement. To navigate this transformative space, filmmakers and industry professionals need to understand the following essential Web3 terms:

1. Blockchain

Definition: A decentralized, immutable ledger that records transactions across multiple computers in a secure and transparent manner.

Relevance to Film3: Blockchain ensures transparency in funding, royalties, and ownership. Filmmakers can use it to establish trust with investors and audiences while ensuring fair profit-sharing.

2. Smart Contracts

Definition: Self-executing contracts with the terms of the agreement directly written into code.

Relevance to Film3: These automate royalty payments, distribution agreements, and licensing. For instance, a smart contract could automatically pay a film's contributors every time a film is streamed.

3. NFTs (Non-Fungible Tokens)

Definition: Unique digital assets stored on a blockchain, representing ownership of specific items such as art, music, or film rights.

Relevance to Film3: NFTs allow filmmakers to tokenize their projects, offering exclusive content, collectibles, or even fractional ownership to fans and investors. This opens new revenue streams and deepens fan engagement.

4. DAO (Decentralized Autonomous Organization)

Definition: An organization governed by smart contracts and blockchain-based voting, allowing collective decision-making without centralized control.

Relevance to Film3: DAOs can fund films or create communities where stakeholders vote on production decisions, fostering democratization in filmmaking.

5. Tokenization

Definition: The process of converting rights or assets into blockchain tokens that can be traded or sold.

Relevance to Film3: Filmmakers can tokenize revenue streams or intellectual property, giving fans and backers a stake in a film's success.

6. Metaverse

Definition: A virtual, immersive environment where users can interact, socialize, and transact.

Relevance to Film3: Filmmakers can premiere movies, host Q&A sessions, or create interactive experiences in the metaverse, offering unique ways to engage audiences.

7. Decentralization

Definition: A system where control is distributed across a network rather than centralized in one authority.

Relevance to Film3: Decentralization empowers creators to bypass traditional gatekeepers like studios, enabling direct collaboration with audiences and investors.

8. Web3 Wallets

Definition: Digital wallets that store cryptocurrencies and NFTs, allowing users to interact with blockchain-based applications.

Relevance to Film3: Audiences and investors need wallets to purchase NFTs or tokens related to a film, while filmmakers can use wallets to manage assets and earnings.

9. Interoperability

Definition: The ability of different blockchain systems and platforms to work together seamlessly.

Relevance to Film3: Ensures that digital assets like NFTs can move freely across platforms, enhancing accessibility for global audiences.

10. Crowdfunding 3.0

Definition: A blockchain-based model where creators raise funds using token sales or NFTs.

Relevance to Film3: This modernized crowdfunding model enables fans to become stakeholders and participate in a film's journey, fostering deeper loyalty.

11. Immutable Records

Definition: Data that cannot be altered once recorded on the blockchain.

Relevance to Film3: Guarantees transparency in production and distribution, allowing stakeholders to verify financial and contractual records.

12. Gas Fees

Definition: Transaction fees required to process operations on a blockchain network.

Relevance to Film3: Understanding gas fees is crucial for managing costs when minting NFTs or using smart contracts for film-related transactions.

13. Layer 2 Solutions

Definition: Secondary frameworks or protocols built on top of blockchains to improve scalability and reduce transaction costs.

Relevance to Film3: Helps filmmakers manage blockchain-related expenses, making tools like NFTs and smart contracts more accessible.

14. P2P (Peer-to-Peer)

Definition: Direct interactions between two parties without intermediaries, enabled by blockchain technology.

Relevance to Film3: Allows creators to distribute films or engage audiences without relying on platforms like streaming giants.

15. Proof of Stake (PoS)

Definition: A blockchain consensus mechanism where validators are chosen based on the number of tokens they hold and are willing to stake.

Relevance to Film3: Many eco-conscious creators prefer PoS-based blockchains due to their lower energy consumption compared to Proof of Work (PoW).

16. Web3 Identity

Definition: A blockchain-based digital identity that allows users to own and control their data across platforms.

Relevance to Film3: Fans and creators can maintain their privacy while engaging in tokenized ecosystems and metaverse experiences.

17. IPFS (InterPlanetary File System)

Definition: A decentralized file storage protocol that distributes content across a network of nodes.

Relevance to Film3: Filmmakers can store large files (like film reels or promotional materials) securely and cost-effectively, ensuring content longevity.

18. Governance Tokens

Definition: Tokens that grant holders voting rights within a DAO or other blockchain project.

Relevance to Film3: These can empower backers to influence decisions about a film's marketing, casting, or distribution.

19. ZK-Proofs (Zero-Knowledge Proofs)

Definition: A cryptographic method allowing one party to prove they know something without revealing the actual information.

Relevance to Film3: Ensures privacy for sensitive transactions, like revenue sharing or proprietary content access.

20. dApps (Decentralized Applications)

Definition: Applications built on blockchain technology that operate without central control.

Relevance to Film3: Filmmakers can use dApps for transparent revenue management, content distribution, or fan interaction.

Why Understanding These Terms Matters

The Film3 movement is more than just a technological shift—it's a paradigm change in how films are created, funded, and consumed. By embracing these terms and their associated technologies, filmmakers can:

- **Secure funding and ownership:** Leverage blockchain transparency to attract investors.
- **Engage audiences:** Foster a sense of community through NFTs, DAOs, and metaverse interactions.
- **Streamline operations:** Automate complex processes like royalty payments with smart contracts.
- **Remain competitive:** Adapt to the evolving expectations of a digitally native, Web3-savvy audience.

As the Film3 ecosystem matures, understanding and applying these concepts will be pivotal to thriving in the next generation of filmmaking.

APPENDIX

RESOURCES AND TOOLS

Web3 and Film3 offer groundbreaking opportunities for filmmakers to redefine how they fund and create their projects. By leveraging these tools, you can build a direct, meaningful connection with your audience, ensure creative freedom, and tap into a global network of supporters. The future of filmmaking is decentralized—and it's yours to shape.

Below is a curated list of key resources and tools to help you begin your Film3 journey, including educational materials, platforms, and agencies that specialize in Web3 support services, among them, **Bespoke Media Marketing**. These resources will help filmmakers navigate the evolving landscape of Web3, from funding and NFT issuance to community building, marketing, and distribution.

1. Educational & Community Resources

1 Film3 Telegram & Discord Groups

- **Film3 Squad (Discord)**
- **NFT Filmmakers (Telegram)**
- **Film.io Community (Discord)**
- These communities allow you to share knowledge, ask questions, and discover current Film3 initiatives.

2. Web3-Focused News Outlets & Blogs

- **Decrypt** – Covers blockchain, NFTs, and decentralized platforms, often discussing Film3 developments.
- **CoinDesk** – Offers industry updates, trend insights, and in-depth analyses on the crypto and NFT market.
- **NFT Now** – Features stories on NFTs, creators, and Web3 art, sometimes including film-based projects.

3. Online Courses & Educational Platforms

- **Cryptomintary.com** – Provides free and paid blockchain fundamentals courses.
- **Odyssey DAO** – Offers workshops on Web3, NFTs, tokenomics, and decentralized governance.
- **Film3-related Panels & Talks (YouTube)** – Many blockchain conferences and NFT summits host dedicated Film3 panels; these often get uploaded to YouTube.

2. Tools & Platforms for Film3

1. NFT Creation & Distribution Platforms

- **OpenSea** – The largest NFT marketplace; offers tools for minting and selling NFTs (including film clips, collectible frames, behind-the-scenes material, etc.).
- **Rarible** – A user-friendly marketplace that allows creators to set custom royalty splits.
- **Mintable** – A no-code solution for NFT minting with flexible royalty options.

2. Funding & Crowdfunding

- **Mirror** – A Web3 publishing and crowdfunding platform where filmmakers can create tokenized crowdfunds (via NFTs or tokens) for their projects.
- **Film.io** – A decentralized filmmaking platform empowering creators to pitch and fund projects through token-based governance.
- **Decentralized Pictures** – A nonprofit platform leveraging blockchain to financially support independent filmmakers.

3. Storage & Content Hosting

- **Arweave** – A permanent, decentralized data storage protocol that can store entire films or related digital assets on-chain.
- **IPFS + Pinata** – IPFS is a peer-to-peer hypermedia protocol for decentralized content storage; Pinata adds easier file management and pinning.
- **Filecoin** – Decentralized storage network that rewards users for storing files in a trustless manner—suitable for large film files.

4. Smart Contract Creation & Management

- **Alchemy** – A powerful development platform that offers tools to build, monitor, and manage dApps on Ethereum and other chains.
- **Thirdweb** – Provides pre-built smart contract templates, making it simpler for filmmakers to deploy NFTs or token contracts without extensive coding.

- **Remix IDE** – A browser-based IDE for writing, compiling, and deploying Solidity smart contracts.

5. Token Gating & Membership Communities

- **Guild.xyz** – Enables token-based membership functionalities on Discord and Telegram, helpful for private screenings and exclusive behind-the-scenes.
- **Collab.Land** – A bot that manages token-based community access on Discord or Telegram, frequently used by NFT-based film communities.

3. Agencies Specializing in Web3 Support

A. Bespoke Media Marketing

- **Services**: Bespoke marketing strategies, brand strategy, go-to-market plans for blockchain-based companies and filmmakers, influencer and brand partnerships, and community building for Film3/NFT projects.
- **Why They're Notable**: They combine traditional entertainment marketing knowledge with blockchain-savvy strategies, helping to position Film3 projects in both Web3 and mainstream channels.

B. Serotonin

- **Services**: Brand strategy, product marketing, and go-to-market plans for blockchain-based companies.
- **Film3 Relevance**: Experienced in launching blockchain and NFT initiatives; can help with storytelling, PR, and community engagements relevant to film projects.

C. MarketAcross

- **Services**: Crypto-focused content marketing, PR, and influencer outreach.
- **Film3 Relevance**: Known for working with major blockchain projects; can guide independent film creators to grow awareness in Web3-centric communities and publications.

D. EAK Digital

- **Services**: PR, event organization, and marketing for crypto and blockchain companies.
- **Film3 Relevance**: Has helped launch various NFT and metaverse brands, giving them a strong network to help film projects gain traction.

E. Crowdcreate

- **Services**: Community management, brand strategy, and influencer marketing across emerging technologies.
- **Film3 Relevance**: Offers robust community-building strategies, ensuring filmmakers can connect effectively with early adopters and crypto-enthusiasts.

4. Marketing & Promotion Tools

1. Social Media & Community Engagement

- **Discord** – Use custom channels, events, and gated content for your project's fanbase and backers.
- **Reddit (r/CryptoCurrency, r/NFT, r/Filmmakers)** – Great for exposure to a broader crypto audience and niche film communities.

2. Video & Livestream Platforms

- **YouTube / Twitch** – For hosting live Q&A sessions, teaser drops, and behind-the-scenes.

3. Press Release Distribution

- **PR Newswire / Business Wire** – Traditional channels that can help broadcast major announcements beyond the Web3 sphere.
- **CoinTelegraph, NFT News Today** – Crypto-focused outlets that highlight noteworthy blockchain projects, including Film3.

5. Networking & Events

1. NFT & Blockchain Conferences

- **NFT NYC** – Prominent NFT-focused event with dedicated film, entertainment, and IP segments.
- **ETHGlobal** – Ethereum hackathons and conferences, often featuring creative uses of NFTs and decentralized storytelling.
- **Consensus by CoinDesk** – Major crypto event covering various use cases, including Film3.

2. Film-Focused Festivals & Events

- **Cannes Next** – An innovation hub at Cannes Film Festival exploring new forms of finance and distribution, including blockchain.
- **Sundance's New Frontier** – Features emerging media, including virtual reality, NFTs, and interactive storytelling; a potential space for Film3.

- **Raindance Film Festival** – Known for focusing on independent film and new distribution technologies, occasionally featuring Web3 events and panels.

6. Additional Tips

- **Start Small**: Experiment with short films or behind-the-scenes footage as NFTs to build awareness and gain hands-on Web3 experience.
- **Leverage DAOs**: Creating or participating in a Decentralized Autonomous Organization (DAO) can help with collaborative funding, decision-making, and governance.
- **Focus on Community**: The key to success in Film3 lies in an engaged and supportive community. Keep them involved in the creative process.
- **Legal & Compliance**: Consult a legal professional with Web3 expertise to ensure compliance with securities laws, IP rights, and film distribution regulations.

Stepping into the Film3 space requires a blend of filmmaking expertise, blockchain savvy, and a strong support network. By leveraging the platforms, tools, and agencies listed above—you can effectively fund, market, and distribute your film projects in this emerging decentralized ecosystem. Embrace the power of community-driven financing, transparent IP rights, and innovative distribution models to make your Film3 endeavor a lasting success.

AFTERWORD

The journey of Web3 film funding is just beginning. In **Volume 1**, we laid the foundation—exploring the power of decentralized financing, community-driven storytelling, and the ways blockchain is disrupting traditional film models. We uncovered how NFTs, DAOs, and crowdfunding are transforming access to capital, giving filmmakers and audiences a new stake in the creative process. This volume has equipped you with the fundamental strategies to navigate this evolving space, empowering you to think beyond traditional funding models and embrace the opportunities of Web3.

But understanding the possibilities is only the first step.

In **Volume 2**, we take a deeper dive into the technical and strategic mechanisms that make Web3 film funding not just possible but scalable. We will explore the intricate world of **smart contracts**, where automated agreements remove intermediaries and ensure transparency in revenue sharing. We'll break down **tokenization**, demonstrating how digital assets can represent everything from ownership in a film to exclusive audience perks. And we'll examine how **on-chain governance** through DAOs enables filmmakers and supporters to make collective creative and financial decisions.

Web3 is not a passing trend—it is an evolution in how we create, distribute, and experience cinema. As the industry continues to shift, those who adapt and innovate will be the ones who shape its future. If **Volume 1** was the blueprint, **Volume 2** is the toolkit, providing you with the deeper knowledge to execute, scale, and sustain Web3-driven film projects.

The future of film is decentralized, and it is being written by those bold enough to embrace it.

See you in **Volume 2**.

Let's Create the Future Together!

ABOUT THE AUTHOR

Photo by Patrick King

Nea Simone is a multi-award-winning, nationally recognized author, storyteller, and visionary in media and entertainment. Her debut novel, *Reaching Back*, has been reprinted over 16 times since its release, captivating readers worldwide and cementing her place as a literary voice of depth and authenticity. The trilogy—*Reaching Back, ReBorn*, and *The Substance of Things*—has garnered critical acclaim, with *ReBorn* earning a coveted spot on the *New York Times* bestseller list.

Recognized by *Borders Bookstore* and *Barnes & Noble* as *An Original Voice of Fiction*, Nea has been honored with the *Reader's Choice in Fiction* award and featured in NBC's "Good People" segment. Her novels, known for their profound storytelling and cultural resonance, are sold in over 20 countries and have inspired countless readers.

Beyond literature, Nea Simone is a pioneer in film marketing, entrepreneurship, and Web3 innovation. As the founder of **Bespoke Media Marketing**, she has created groundbreaking campaigns connecting multicultural audiences with independent films, working with entertainment and technology industry giants. She is also the visionary behind **Melanated**

Studios, the first Black woman-owned blockchain technology company dedicated to the NFT and Metaverse space.

Nea's expertise extends to the global stage, where she has spoken at Cannes, NFT ATLANTA CON, and the African Cinema Summit. She has been featured in *Variety, CoinTelegraph, NFT Plazas*, and *Yahoo Entertainment* and has served as a judge for major Web3 hackathons. Her thought leadership in blockchain technology, digital storytelling, and audience engagement continues to shape the future of entertainment.

When she's not writing or building industry-shifting strategies, Nea is deeply engaged in empowering underrepresented voices, pushing the boundaries of storytelling, and finding innovative ways to connect stories with the world.

For more information, visit:
neasimone.com | bespokemediamarketing.com

Follow on social media:
@bespokemediamarketing @neasimonewrites

www.ingramcontent.com/pod-product-compliance
Lightning Source LLC
LaVergne TN
LVHW022224200926
848481LV00006B/2111
* 9 7 8 0 9 8 5 8 8 3 3 9 3 *